AF601493

Synthesis Lectures on Engineering, Science, and Technology

The focus of this series is general topics, and applications about, and for, engineers and scientists on a wide array of applications, methods and advances. Most titles cover subjects such as professional development, education, and study skills, as well as basic introductory undergraduate material and other topics appropriate for a broader and less technical audience.

Sujaul Chowdhury

Introduction to Electronics

Sujaul Chowdhury
Department of Physics
Shahjalal University of Science
and Technology
Sylhet, Bangladesh

ISSN 2690-0300 ISSN 2690-0327 (electronic)
Synthesis Lectures on Engineering, Science, and Technology
ISBN 978-3-032-03448-9 ISBN 978-3-032-03449-6 (eBook)
https://doi.org/10.1007/978-3-032-03449-6

This Springer imprint is published by the registered company Springer Nature Switzerland AG
The registered company address is: Gewerbestrasse 11, 6330 Cham, Switzerland

Preface

A course titled *Introduction to Electronics* is taught in all universities, but it is a pity that there is no available book that can be adopted for classroom use. Moreover, authors of most available books on introductory electronics are engineers rather than physicists, and hence, there is lack of clarity on physics in these books. Students of physics find these books uncomfortable to go through. These books contain large volumes of texts and are unsuitable for one-semester course.

This book titled *Introduction to Electronics* is based on my lecture notes for a one semester course that I conducted for undergraduates of physics major in their 2nd year in Department of Physics, Shahjalal University of Science and Technology, Sylhet, Bangladesh. It proved difficult to prepare coherent and ready-to-go lecture notes for the course using available books.

This book contains concise, coherent, and ready-to-go lecture notes of the course and is ideal for a one semester course. Thus, this book is expected to be a great contribution to teaching and learning. Pre-requisite for the course is a course on Electricity and Magnetism that is usually taught in the 1st year.

There are 7 chapters:

Chapter 1 Introductory Topics

This chapter covers introductory topics that are necessary to grasp the remaining chapters of this book. Starting with an understanding of Ohm's law, it covers use of series resistor circuit as potential divider, parallel resistor circuit as current divider, and reduction of series-parallel combination circuit. Understanding of galvanometer, ammeter, and voltmeter is also covered. Wheatstone bridge and meter bridge are also introduced.

Chapter 2 DC Circuit Analysis

This chapter covers various methods of analyzing direct current (dc) circuits. The chapter starts with Kirchhoff's current and voltage laws and their use in branch current method (called standard method), mesh loop method and nodal analysis method of analyzing dc circuits. Calculations for inter-conversion between Del and Y circuit configurations are presented in complete detail. Superposition theorem, Thevenin's theorem, Norton's theorem and maximum power transfer theorem are covered in some detail. Transient current in *RC* series dc circuit is also analyzed.

Chapter 3 AC Circuit Analysis

This chapter covers various methods of analyzing and utilizing alternating current (ac) circuits. Starting with an introduction to alternating current, responses of resistor, capacitor and inductor as well as those of series combinations of these have been calculated using trigonometric calculations, phasor and complex impedance methods. It has been shown that these circuits can be utilized as low-pass, high-pass and band-pass filters as well as phase shifters besides as differentiating and integrating circuits.

Chapter 4 p–n Junction

This chapter deals with p-n junction and uses of it in detail. The chapter starts with energy band and energy gap of semiconductor crystals like Silicon and n-type and p-type doping of the semiconductor crystal. It then covers energy band model of p-n junction. Use of p-n junction in half-wave and full-wave bridge rectification and in voltage regulation is covered besides a nice coverage of light emitting diode and photodiode using energy band models.

Chapter 5 Bipolar Junction Transistor (BJT)

This chapter deals with Bipolar Junction Transistor (BJT). It covers structure and band models of and currents through BJT. Current-voltage characteristics and biasing of, and ac voltage amplification by BJT are also covered.

Chapter 6 Junction Field Effect Transistor (JFET)

This chapter deals with n-channel junction field effect transistor (n-JFET). The chapter covers structure and biasing of and ac voltage amplification by n-JFET.

Chapter 7 Metal Oxide Semiconductor Field Effect Transistor (MOSFET)

This chapter deals with both depletion type and enhancement type n-channel metal oxide semiconductor field effect transistor (MOSFET). The chapter covers structure and biasing of and ac voltage amplification by MOSFET.

Sylhet, Bangladesh
2025

Sujaul Chowdhury

Contents

Introductory Topics

1

1.1 Electric Current

See Figs. 1.1 and 1.2. Free electrons called conduction electrons in metal (wire) are in random motion with large speed. Number of these electrons crossing a plane perpendicular to axis of the wire from left to right is the same as number of electrons crossing the plane from right to left in any given time interval. Thus no net current is found in absence of a battery.

See Figs. 1.1 and 1.2. If we connect a battery, a small drift velocity v_d gets superimposed on the random velocity. A free electron travels $AB'C'D'$ instead of $ABCD$ in time interval Δt. Drift velocity is average speed with which charged particles like electrons move in a conductor when an electric field is applied. The velocity is given by $v_d = DD'/\Delta t$ which gets superimposed on random velocity and we get a net measured current (across the plane).

The drift velocity is very small, typically ~ 10^{-5} m/s. But the number of free electrons in metal such as Cu is very large ~ $10^{22}/\text{cm}^3$. As such we get a large current due to small battery bias. The current is $I = q/t$ or dq/dt. Here q is net charge passing across the plane in time t. Direction of the current I is taken opposite to direction of flow of electrons for historical reason. Unit of current is Ampere (A).

1.2 Potential Difference or Voltage

Work or energy needed to accelerate a free electron and move it from one point to another in a material is called potential difference of electron between the two points. Potential difference V is quantified as $V = W/q$ which is the work W per unit charge. Unit of V is

S. Chowdhury, *Introduction to Electronics*, Synthesis Lectures on Engineering, Science, and Technology, https://doi.org/10.1007/978-3-032-03449-6_1

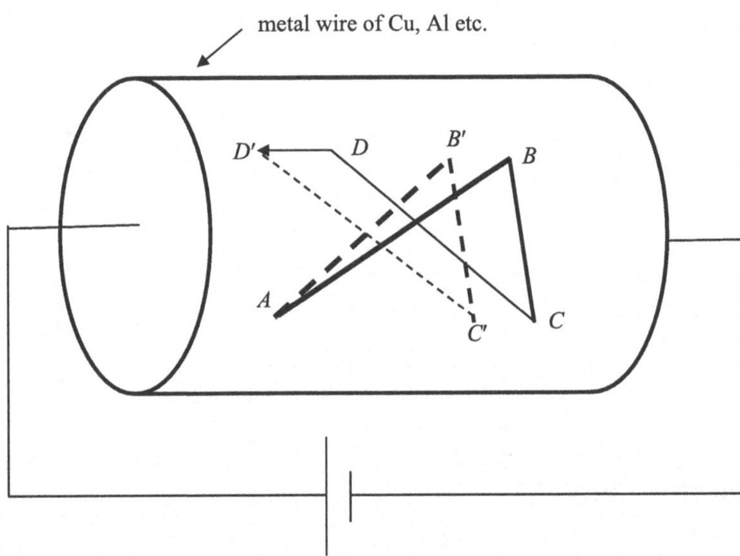

Fig. 1.1 Showing random motion of an electron which travels from A to D along AB, BC, CD in a time interval Δt in absence of a battery. In presence of battery, the electron travels from A to D' along AB', $B'C'$, $C'D'$ in the same time interval Δt. Thus a net displacement DD' occurs in the said time interval. $DD'/\Delta t$ is called drift velocity v_d which is very small compared with the random velocity

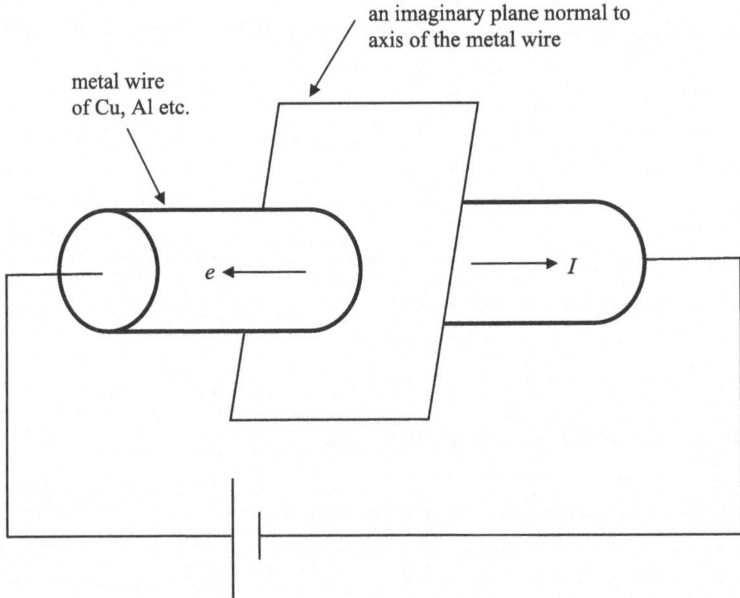

Fig. 1.2 To help understand electric current. Direction of flow of electron e and direction of conventional current I are indicated

Volt (V). A battery can do this work and potential difference that the battery can provide is called electromotive force or emf. The emf is work, not force.

1.3 Resistance

Free electrons called conduction electrons encounter resistance against flow in material. This is due to multiple collisions. Resistance R is related to resistivity ρ as $R = \rho L/A$ where L is length and A is area of cross section of the material through which current is trying to flow due to battery bias. Unit of resistance R is Ohm (Ω). $K = 1/R$ is called conductance unit of which is mho or Siemen (S).

1.4 Ohm's Law and Joule's Law

To maintain a larger current I in a given material, more energy or larger potential difference V is needed. We have $V \propto I$ or,

$$V = RI \tag{1.1}$$

called Ohm's law. Here R is proportionality constant called resistance.

Kinetic energy of electrons provided by battery is dissipated by resistance and the energy converts to heat at the rate

$$P = \frac{dW}{dt} = \frac{d}{dt}(Vq) = V\frac{dq}{dt} = VI = RI^2 = \frac{V^2}{R} \tag{1.2}$$

using $V = W/q$, $I = dq/dt$ and $V = R\,I$.

1.5 Battery

In battery, chemical reactions maintain excess of free electrons at one terminal and lack of free electrons at the other terminal. As such, if we connect a battery with a resistor of resistance R, a potential difference or voltage drop V or E called emf will be maintained across R and a current $I = V/R$ or $I = E/R$ will flow. See Fig. 1.3.

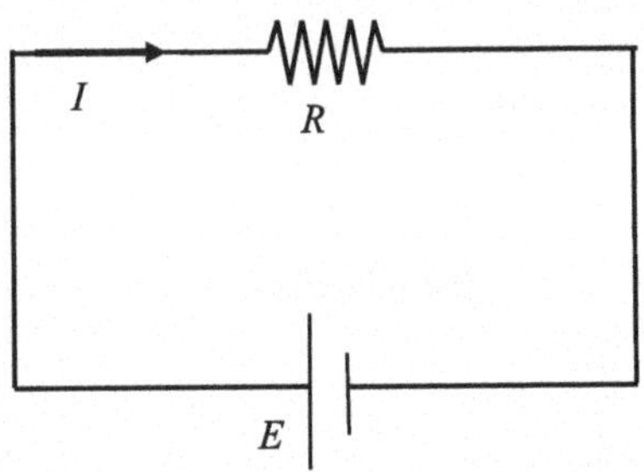

Fig. 1.3 An electrical circuit containing a battery of emf E and a resistor of resistance R. A current $I = E/R$ passes through the resistor

1.6 Series Circuit: The Voltage Divider

See Fig. 1.4. We have 3 resistors connected so that same current I passes through each of these. Such a circuit is called *series circuit*. Potential difference across the resistors of resistances R_1, R_2 and R_3 are $V_1 = R_1 I$, $V_2 = R_2 I$ and $V_3 = R_3 I$. Therefore $E = V_1 + V_2 + V_3 = (R_1 + R_2 + R_3)\, I = R_{\mathrm{eq}}\, I$ where $R_{\mathrm{eq}} = R_1 + R_2 + R_3$ is called *equivalent resistance*. This is equivalent resistance law for series circuit. The voltage E divides into 3 parts, $V_1 = R_1 I$, $V_2 = R_2 I$, $V_3 = R_3 I$ in the ratio

$$V_1 : V_2 : V_3 = R_1 : R_2 : R_3.$$

If we have 2 resistors of resistances R_1 and R_2 in series with a battery of emf E, we have $E = V_1 + V_2 = (R_1 + R_2)\, I = R_{\mathrm{eq}}\, I$ where

$$R_{\mathrm{eq}} = R_1 + R_2 \tag{1.3}$$

is the *equivalent resistance*. The voltage E divides into 2 parts, $V_1 = R_1 I$ and $V_2 = R_2 I$ in the ratio

$$V_1 : V_2 = R_1 : R_2 \tag{1.4}$$

Thus series circuit works as voltage divider.

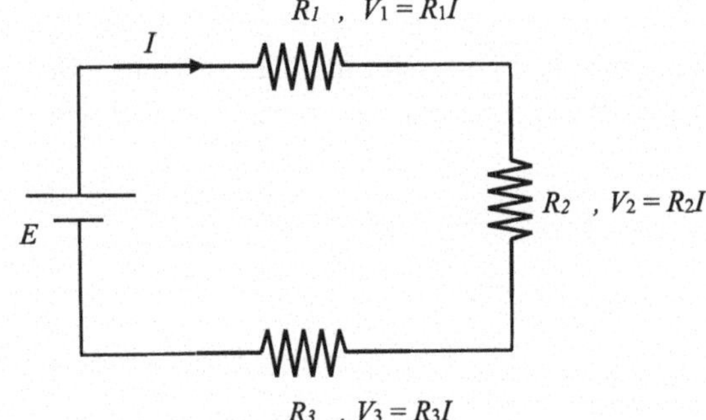

Fig. 1.4 Showing a series circuit containing 3 resistors of resistances R_1, R_2, R_3 and a battery of emf E. Same current I passes through each of the 3 resistors

1.7 Parallel Circuit: The Current Divider

See Fig. 1.5. We have 3 resistors connected so that potential difference across each of these are equal or same, which is E. Currents through the resistors are determined by Ohm's law as $I_1 = \frac{E}{R_1}$, $I_2 = \frac{E}{R_2}$ and $I_3 = \frac{E}{R_3}$.

We have $I = I_1 + I_2 + I_3 = E/R_1 + E/R_2 + E/R_3 = E(1/R_1 + 1/R_2 + 1/R_3)$ or,

$$E/R_{eq} = E(1/R_1 + 1/R_2 + 1/R_3)$$

Thus

$$1/R_{eq} = 1/R_1 + 1/R_2 + 1/R_3$$

where R_{eq} is called *equivalent resistance.* This is equivalent resistance law for parallel circuit.

See Fig. 1.6. For a parallel circuit containing 2 resistors of resistance R_1 and R_2, equivalent resistance is R_{eq} given by

$$1/R_{eq} = 1/R_1 + 1/R_2 \tag{1.5}$$

Thus

$$R_{eq} = \frac{R_1 R_2}{R_1 + R_2} \tag{1.6}$$

We have

$$R_{eq} = \left(\frac{R_1}{R_1 + R_2}\right) R_2 = \left(\frac{R_2}{R_1 + R_2}\right) R_1$$

which means R_{eq} is less than both R_1 and R_2.

We have

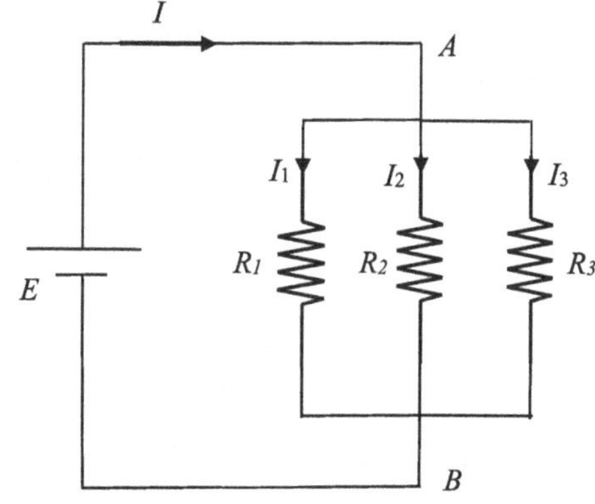

Fig. 1.5 Showing a parallel circuit containing 3 resistors of resistances R_1, R_2, R_3 and a battery of emf E. The main current I divides into 3 parts and these currents pass through the 3 resistors

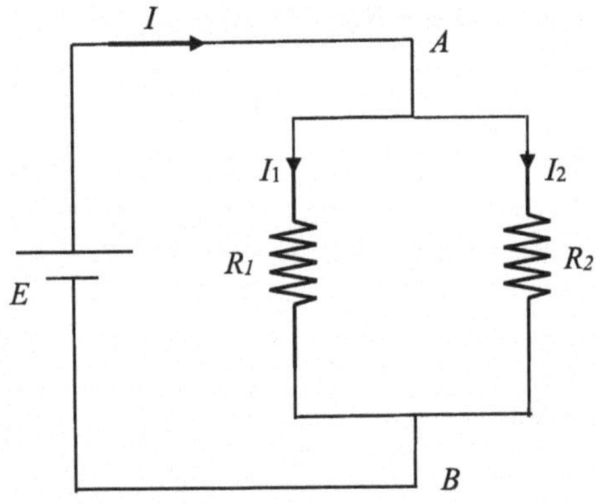

Fig. 1.6 Showing a parallel circuit containing 2 resistors of resistance R_1 and R_2 and a battery of emf E. The main current I divides into 2 parts and these pass through the 2 resistors

$$I = I_1 + I_2$$

where

$$I_1 = E/R_1$$

and

$$I_2 = E/R_2$$

because potential difference is the same across both R_1 and R_2 and is equal to E. Thus the current I divides into 2 parts I_1 and I_2 in the ratio

$$\frac{I_1}{I_2} = \frac{R_2}{R_1} \tag{1.7}$$

called current division rule. Thus we have

$$I_1 = \left(\frac{R_2}{R_1 + R_2}\right)I \tag{1.8}$$

and

$$I_2 = \left(\frac{R_1}{R_1 + R_2}\right)I \tag{1.9}$$

1.8 Series–Parallel Combination: Circuit Reduction Using Equivalent Resistance Laws

Let us carry out reduction of the circuit shown in Fig. 1.7a using equivalent resistance laws. Since R_5 and R_6 are in parallel combination, their equivalent resistance is $R_{eq1} = \frac{R_5 R_6}{R_5+R_6}$. Now R_4 and R_{eq1} are in series combination; their equivalent resistance is $R_{\text{eq2}} = R_4 + R_{\text{eq1}} = R_4 + \frac{R_5 R_6}{R_5+R_6}$. Now R_{eq2} is in parallel combination with R_3; their equivalent resistance is $R_{eq3} = \frac{R_3 R_{eq2}}{R_3+R_{eq2}} = \frac{R_3\left(R_4+\frac{R_5R_6}{R_5+R_6}\right)}{R_3+\left(R_4+\frac{R_5R_6}{R_5+R_6}\right)}$. Finally, R_1, R_{eq3} and R_2 are in series combination. As such, total resistance faced by the battery is

$$R_{eq} = R_1 + R_{eq3} + R_2 = R_1 + \frac{R_3\left(R_4 + \frac{R_5R_6}{R_5+R_6}\right)}{R_3 + \left(R_4 + \frac{R_5R_6}{R_5+R_6}\right)} + R_2$$

and $I = E/R_{\text{eq}}$. These are what Fig. 1.7b is trying to portray.

See Fig. 1.7c. If we now wish to know I_3, we note that main current I divides into 2 parts, namely I_3 and I_{eq2}. Now current division rule gives $I_3 = \left(\frac{R_{eq2}}{R_3+R_{eq2}}\right)I$ or, $I_3 = \left(\frac{R_4+\frac{R_5R_6}{R_5+R_6}}{R_3+R_4+\frac{R_5R_6}{R_5+R_6}}\right)I$. Thus we have got familiar with use of equivalent resistance laws in reduction of complex series–parallel combination circuit.

1.9 Galvanometer, Ammeter and Voltmeter

Galvanometer is a device in which deflection θ of the needle is *proportional* to current I passing through it, i.e. $\theta \propto I$. Figure 1.8 shows a galvanometer between a and b in a series circuit. G is resistance of the coil of the galvanometer. Thus galvanometer can *detect* presence of current in an electrical circuit.

Ammeter can be used to *measure* current I in a circuit. If we use a resistor of very small resistance S called *shunt* in parallel combination with galvanometer resistance G, we get what is called an ammeter; see Fig. 1.9 in which there is an ammeter between a and b in series with a resistor and a battery.

(1) Resistance between a and b is $\frac{GS}{G+S} \approx \frac{GS}{G} \approx S$ and it is very small because S is very small.
(2) Resistance between a and b is smaller than S because it is $\frac{GS}{G+S} = \left(\frac{G}{G+S}\right)S$.
(3) If we wish to measure current I passing through the resistor of resistance R in the circuit of Fig. 1.9, we need to connect an ammeter *in series* with the resistor R as in Fig. 1.9.

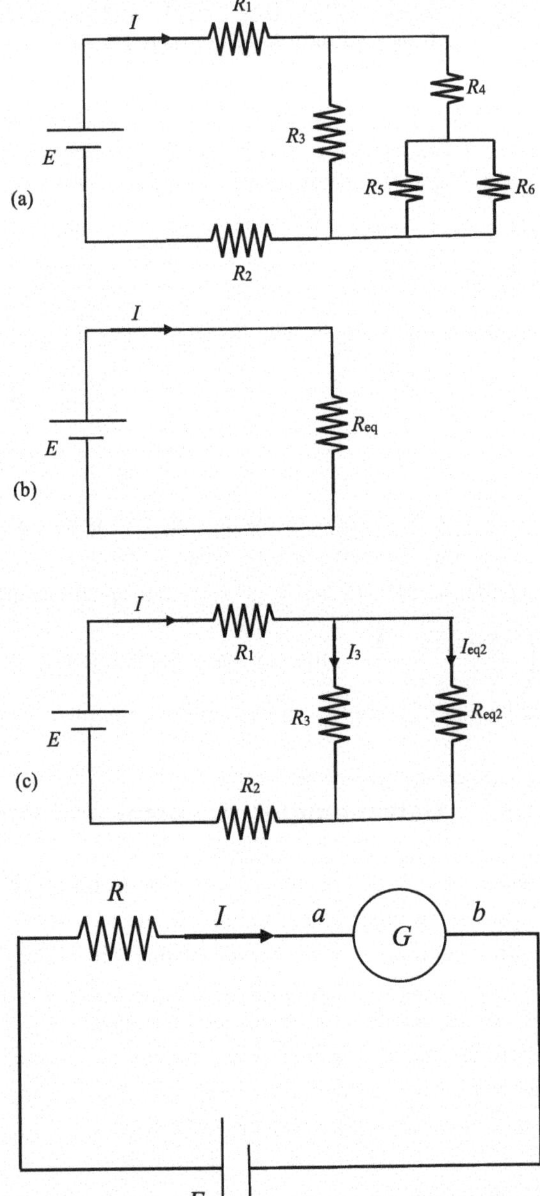

Fig. 1.7 To illustrate circuit reduction using equivalent resistance laws. We wish to reduce circuit of **a** to those of **b**, **c**

Fig. 1.8 Showing a galvanometer in series with a resistor of resistance R and a battery of emf E

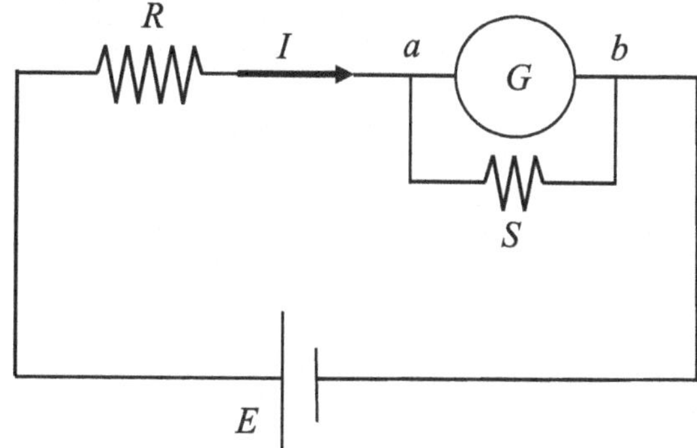

Fig. 1.9 Showing a shunt of resistance S in parallel combination with galvanometer resistance G to get what is called ammeter in series with a resistor of resistance R and a battery of emf E

(4) Since resistance between a and b, which is resistance of the ammeter, is very small, current I passing through R will remain (almost) *unaltered* in the face of insertion of the ammeter in the circuit.
(5) A small fraction of current I passes through the galvanometer coil; most of the current I passes through the shunt S.
(6) The current passing through the galvanometer coil is $I_g = \frac{S}{G+S}I \approx \frac{S}{G}I$ which is very small because $S \ll G$. Hence large current I cannot damage the galvanometer.
(7) The current passing through the shunt is $I_s = \frac{G}{G+S}I \approx \frac{G}{G}I \approx I$ which is almost equal to the total current I.
(8) Since $I_g = \frac{S}{G+S}I$, we still have $I_g \propto I$ and hence if we note galvanometer deflections θ for two or more known currents I, we can calibrate the galvanometer as ammeter and we can measure current using it.

See Fig. 1.10. If we combine a resistor of large resistance R_g in series with a galvanometer, we get what is called voltmeter. We can use it to *detect* and to *measure* potential difference or voltage drop between any two locations of an electrical circuit. In Fig. 1.10, resistance between a and b is $R_g + G$ which is large because R_g is large.

(1) If we wish to measure potential difference or voltage drop across R_2 in the electrical circuit shown in Fig. 1.11, we need to connect a voltmeter in *parallel* combination with R_2 as shown in Fig. 1.11.
(2) Since $R_g + G$ is very large, the current $I_2 \approx I$ and hence voltage drop across R_2 remains (almost) *unchanged* in the face of connecting the voltmeter in the circuit, i.e. $I_2 R_2 \approx IR_2$.

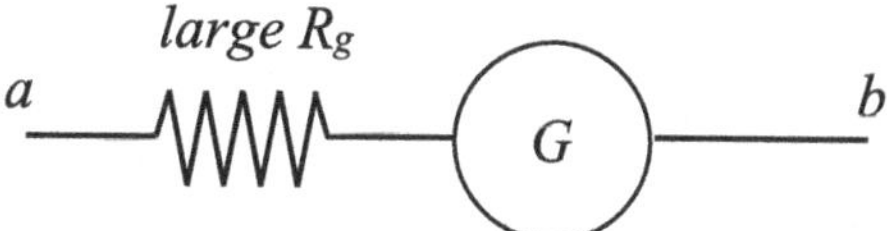

Fig. 1.10 Showing a voltmeter. A resistor of large resistance R_g is connected in series with a galvanometer

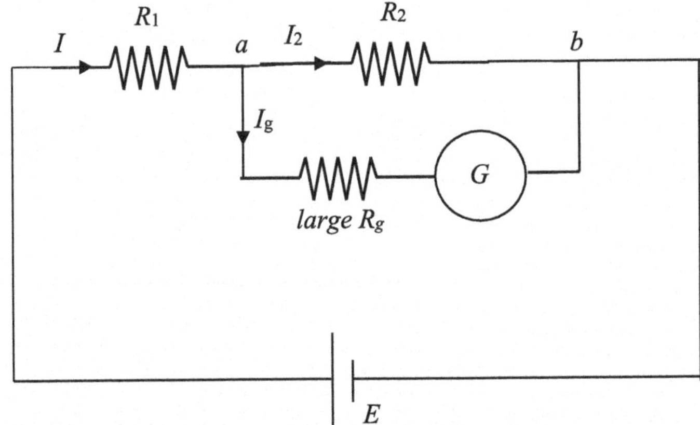

Fig. 1.11 Showing use of voltmeter to measure potential difference or voltage drop across R_2 in the electrical circuit

(3) A very small fraction of current I which is I_g passes through $R_g + G$ and causes galvanometer deflection.
(4) Potential difference between a and b is $V_{ab} = I_g\ (R_g + G)$ as well as $V_{ab} = I\ R_2 \approx I_2\ R_2$.
(5) Hence $I_g\ (R_g + G) = I\ R_2$. Thus $I_g = \frac{R_2 I}{R_g + G}$.
(6) Thus galvanometer deflection θ which is proportional to I_g is also proportional to $R_2\ I$ which is voltage drop across R_2.
(7) Thus if we note the galvanometer deflections θ for two or more known values of (the product) $R_2 I$, we can calibrate the galvanometer scale as voltmeter scale.

1.10 Wheatstone Bridge and Meter Bridge

See Fig. 1.12. The circuit is called Wheatstone bridge. When or if galvanometer deflection is zero, potential difference V_{bc} between the points or locations b and c is zero. Hence we have $V_{ab} = V_{ac}$ or,

$$R_1 I_1 = R_3 I_2$$

Again, since $V_{bc} = 0$, we have $V_{bd} = V_{cd}$ or,

$$R_2 I_1 = R_4 I_2$$

Dividing, we get

$$\frac{R_1}{R_2} = \frac{R_3}{R_4} \tag{1.10}$$

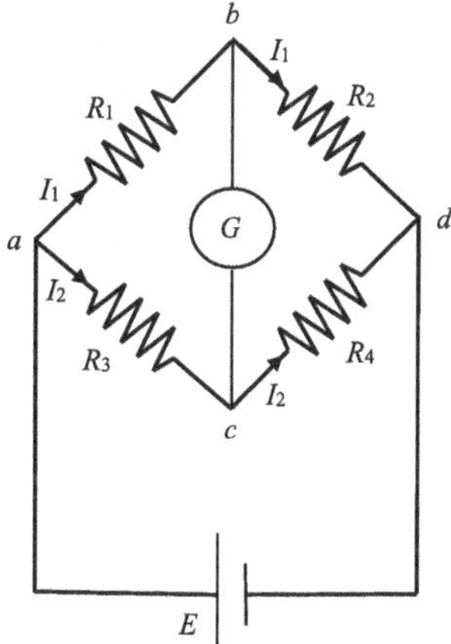

Fig. 1.12 Showing a Wheatstone bridge in equilibrium i.e. with zero galvanometer deflection for which Eq. (1.10): $\frac{R_1}{R_2} = \frac{R_3}{R_4}$ holds

Equation (1.10) is the famous relation for equilibrium or null galvanometer deflection of Wheatstone bridge.

Meter bridge is a user-friendly version of Wheatstone bridge. See Fig. 1.13. It consists of 3 broad metal plates of small resistance, say 0.1 Ω each and a 100 cm long metal wire between *c* and *d*. Point *c* is marked as 0 and point *d* is marked as 100 cm. Hence the name meter bridge.

Figure 1.14 demonstrates how meter bridge is used. A battery is connected between *a* and *b*. Two resistors of resistance R_1 and R_2 are placed in the two gaps. One end of a galvanometer is put at *e* while another end of it is connected with a jockey which can be contacted with the metal wire at *f* anywhere between *c* and *d*. If we call $cf = L$ cm, $fd = (100 - L)$ cm.

If the galvanometer deflection is zero for a particular location of the jockey, we have from Eq. (1.10): $\frac{R_1}{R_2} = \frac{R_3}{R_4}$ that $\frac{R_1}{R_2} = \frac{L\sigma}{(100-L)\sigma}$ which gives

$$\frac{R_1}{R_2} = \frac{L}{100 - L} \tag{1.11}$$

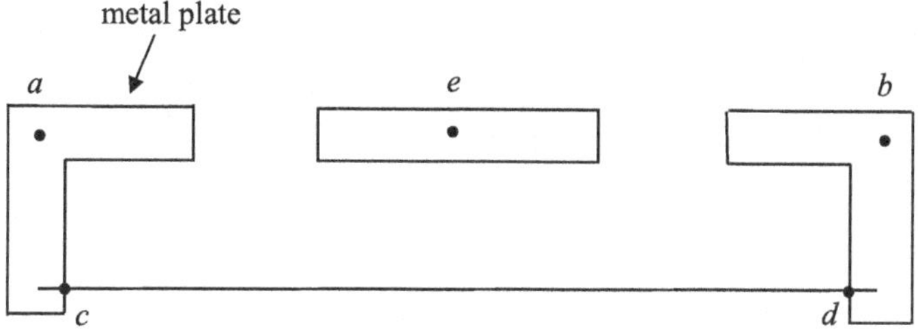

Fig. 1.13 Showing a meter bridge

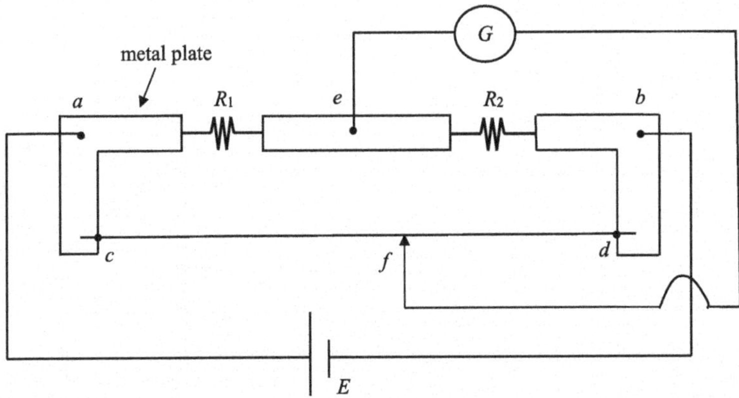

Fig. 1.14 Showing use of meter bridge

Here σ is resistance per unit length of the metal wire. Equation (1.11) holds accurately if

(1) resistance of metal plates is negligible compared with that of R_1 and R_2
(2) resistance between a and c is negligible
(3) resistance between b and d is negligible.

If resistance between a and c which we call R_{ac} is not negligible and if resistance between b and d which we call R_{bd} is not negligible, Eq. (1.10): $\frac{R_1}{R_2} = \frac{R_3}{R_4}$ applied to the meter bridge gives

$$\frac{R_1}{R_2} = \frac{L\sigma + R_{ac}}{(100 - L)\sigma + R_{bd}} \tag{1.12}$$

Equation (1.12) can be written as

$$\frac{R_1}{R_2} = \frac{(L + \lambda_L)\sigma}{((100 - L) + \lambda_R)\sigma} = \frac{L + \lambda_L}{(100 - L) + \lambda_R} \tag{1.13}$$

Here $\lambda_{\mathrm{L}}\sigma$ is resistance between a and c i.e. R_{ac} while $\lambda_{\mathrm{R}}\sigma$ is resistance between b and d i.e. R_{bd}. See Fig. 1.14. λ_{L} and λ_{R} are (equivalent) lengths of meter bridge wire resistances of which are equal to R_{ac} and R_{bd} respectively.

λ_{L} can be determined as follows. See Fig. 1.15. R_{g} is very large so that we have the galvanometer as a voltmeter. The voltmeter deflection $\theta = V_{\mathrm{af}}$ which is voltage between a and f. Thus

$$\begin{aligned} \theta &= V_{\mathrm{af}} = R_{\mathrm{af}}I = (R_{\mathrm{ac}} + R_{\mathrm{cf}})I \\ &= (\lambda_{\mathrm{L}}\sigma + L\sigma)I = \sigma IL + \lambda_{\mathrm{L}}\sigma I \end{aligned}$$

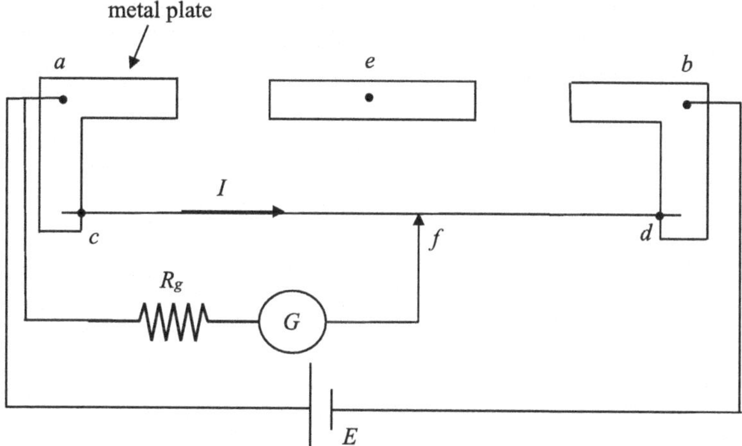

Fig. 1.15 To show how to determine λ_L, the so called end-correction

Thus θ versus L plot is a straight line which intersects L axis at $L = -\lambda_L$ as can be found by putting $\theta = 0$ in the equation $\theta = \sigma I L + \lambda_L \sigma I$. Thus we can determine λ_L. Similarly, we can determine λ_R. In that case the voltmeter in Fig. 1.15 needs to be connected at b rather than at a.

2 DC Circuit Analysis

2.1 Kirchhoff's Current Law and Voltage Law

See Fig. 2.1. a and b are called *nodes* or branch points. We have assigned *branch currents* I_1, I_2, I_3 to various branches of the electrical circuit. At node a, we have $I_1 = I_2 + I_3$ or, $I_1 - I_2 - I_3 = 0$. Thus algebraic sum of currents at node a is zero. For node b, we have $I_2 + I_3 = I_1$ or, $I_2 + I_3 - I_1 = 0$. Thus algebraic sum of currents at node b is also zero. This is Kirchhoff's 1st law or current law. It states that algebraic sum of currents at any node of electrical circuit is zero.

In Fig. 2.1, we have three possible loops, namely *dabcd*, *aefba* and *defcd*. Algebraic sum of potential differences or voltage drops around any such complete loop of a circuit is zero. This is Kirchhoff's 2nd law or voltage law.

As to the loop *dabcd*, we write

$$-R_1I_1 - R_2I_2 - R_3I_2 + E = 0 \tag{2.1}$$

where E is emf of the battery. As to the loop *aefba*, we write

$$-R_4I_3 + R_3I_2 + R_2I_2 = 0 \tag{2.2}$$

As to the loop *defcd*, we write

$$-R_1I_1 - R_4I_3 + E = 0 \tag{2.3}$$

These three equations together with Kirchhoff's current law

$$I_1 - I_2 - I_3 = 0 \tag{2.4}$$

S. Chowdhury, *Introduction to Electronics*, Synthesis Lectures on Engineering, Science, and Technology, https://doi.org/10.1007/978-3-032-03449-6_2

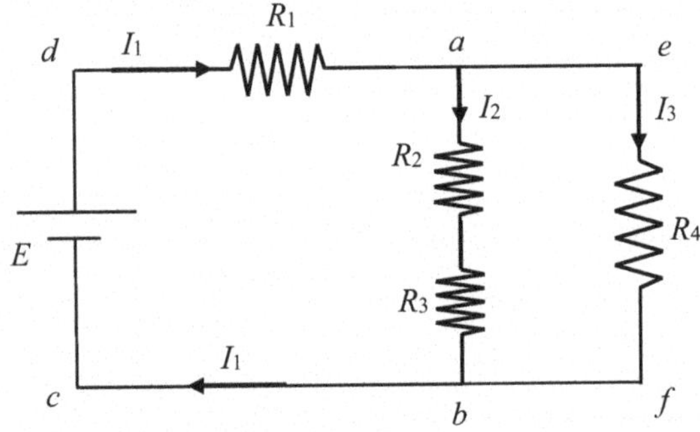

Fig. 2.1 To help introduce Kirchhoff's laws and to describe *branch current method* of circuit analysis

provide us 4 equations in 3 unknown currents; resistances of the resistors are taken as known quantities. Hence we can determine the 3 branch currents using any 3 out of the 4 equations. The method is called *branch current method* of circuit analysis.

2.2 Mesh-Loop Method of Circuit Analysis

In mesh-loop method, we assign currents I_1, I_2, I_3 to the (three) loops (shown in Fig. 2.2) rather than to branches of the circuit. We move around each loop applying Kirchhoff's voltage law taking contributions of all related loops to current through each resistor.

For loop 1, we have

$$E - R_1(I_1 + I_3) - R_2(I_1 - I_2) - R_3(I_1 - I_2) = 0 \tag{2.5}$$

For loop 2, we have

$$-R_4(I_2 + I_3) - R_3(I_2 - I_1) - R_2(I_2 - I_1) = 0 \tag{2.6}$$

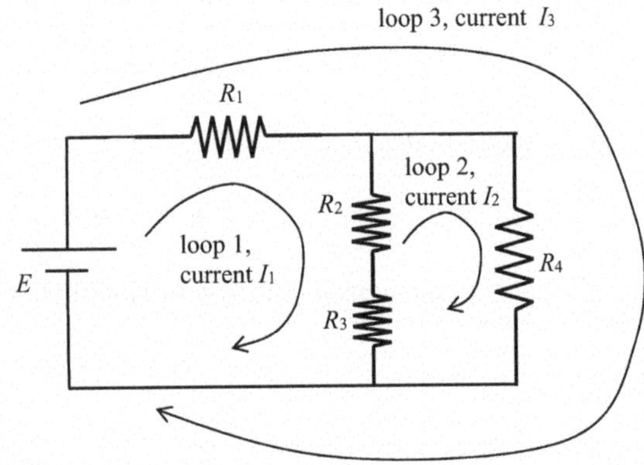

Fig. 2.2 To help describe *mesh-loop method* of circuit analysis

For loop 3, we have

$$E - R_1(I_1 + I_3) - R_4(I_2 + I_3) = 0 \tag{2.7}$$

We can get the loop currents I_1, I_2, I_3 by solving Eqs. (2.5), (2.6) and (2.7). Current through e.g. R_2 comes out as the difference between I_1 and I_2. Current through R_4 comes out as sum of I_2 and I_3. Voltage drop across e.g. R_2 is product of R_2 and $(I_1 \sim I_2)$.

2.3 Nodal Analysis Method of Circuit Analysis

See Fig. 2.3. For node *a*, Kirchhoff's current law gives

$$I_1 = I_2 + I_3$$

which gives

$$\frac{V_c - V_a}{R_1} = \frac{V_a - V_b}{R_2 + R_3} + \frac{V_a - V_b}{R_4}$$

where $V_\mathrm{b} = 0$ and $V_\mathrm{c} = E$. Thus

$$\frac{E}{R_1} = \left(\frac{1}{R_2 + R_3} + \frac{1}{R_4} + \frac{1}{R_1}\right)V_a$$

which gives V_a. Then any branch current can be calculated.

We now present nodal analysis of Wheatstone bridge. See Fig. 2.4. At node *d*, we have

$$I_1 = I_2 + I_g$$

or,

$$\frac{V_a - V_d}{R_1} = \frac{V_d - V_c}{R_2} + \frac{V_d - V_b}{G}$$

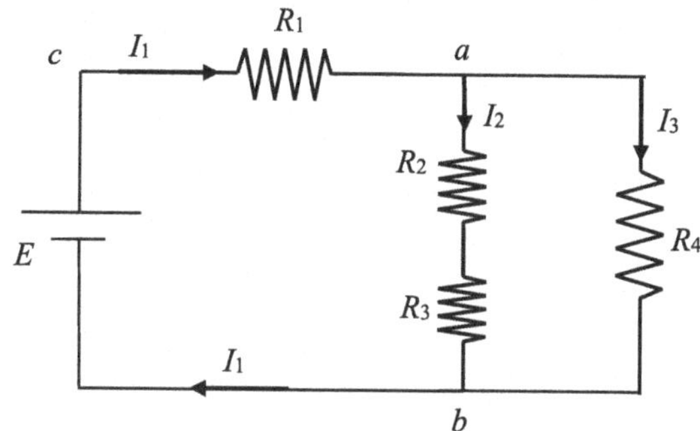

Fig. 2.3 To help describe *nodal analysis method* of circuit analysis

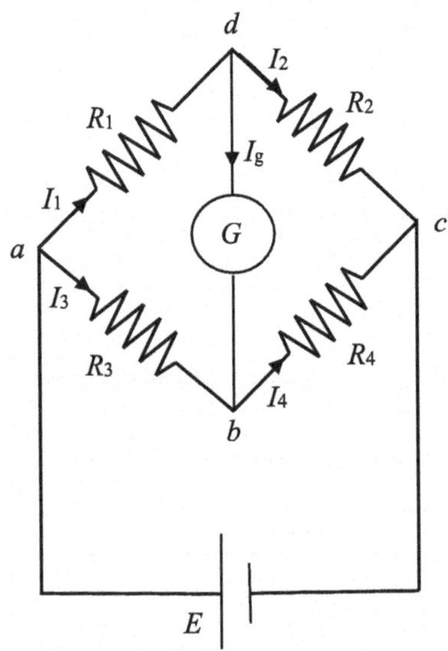

Fig. 2.4 To help analyze Wheatstone bridge using nodal analysis method. $V_a = E$, $V_c = 0$

or,

$$\frac{E - V_d}{R_1} = \frac{V_d}{R_2} + \frac{V_d - V_b}{G} \tag{2.8}$$

At node b, we have

$$I_3 + I_g = I_4$$

or,

$$\frac{V_a - V_b}{R_3} + \frac{V_d - V_b}{G} = \frac{V_b - V_c}{R_4}$$

or,

$$\frac{E - V_b}{R_3} + \frac{V_d - V_b}{G} = \frac{V_b}{R_4} \tag{2.9}$$

We have 2 equations: Eqs. (2.8) and (2.9) in 2 unknowns, namely V_b and V_d. Thus V_b and V_d can be obtained. Thereafter, current through and voltage across any resistor of the Wheatstone bridge can be obtained.

For null or zero deflection of the galvanometer, we have $I_g = 0$ and hence $V_b = V_d$. As such Eqs. (2.8) and (2.9) give

$$\frac{E - V_d}{R_1} = \frac{V_d}{R_2} + 0$$

and

$$\frac{E - V_b}{R_3} + 0 = \frac{V_b}{R_4}$$

respectively. Dividing, we get

$$\frac{R_3}{R_1} = \frac{R_4}{R_2}$$

or,

$$\frac{R_1}{R_2} = \frac{R_3}{R_4} \tag{2.10}$$

which is the famous equation for null galvanometer deflection of Wheatstone bridge.

2.4 Inter-conversion Between Del and Y Circuit Configurations

Inter-conversion between Del and Y configurations of resistors in electrical circuits often proves vital in reduction of complex circuits, particularly when we do not wish to use branch current method, mesh-loop method, and nodal analysis method because of complexity.

For the two circuits to be equivalent, total resistance between any two corresponding terminals must be same. For example, resistance between terminals A and C in the del configuration (Fig. 2.5a) with terminal B open-circuited, which is $R_1 \parallel (R_2 + R_3)$ must be equal to resistance between terminals A and C in Y configuration (Fig. 2.5b) with terminal B open-circuited which is $R_c + R_a$. Thus

$$R_c + R_a = \frac{R_1(R_2 + R_3)}{R_1 + R_2 + R_3} \tag{2.11}$$

Similarly,

$$R_b + R_c = \frac{R_3(R_1 + R_2)}{R_1 + R_2 + R_3} \tag{2.12}$$

$$R_a + R_b = \frac{R_2(R_3 + R_1)}{R_1 + R_2 + R_3} \tag{2.13}$$

Summing Eqs. (2.11), (2.12) and (2.13), we get

$$R_a + R_b + R_c = \frac{R_1R_2 + R_2R_3 + R_3R_1}{R_1 + R_2 + R_3} \tag{2.14}$$

Equation (2.14) – Eq. (2.12) gives

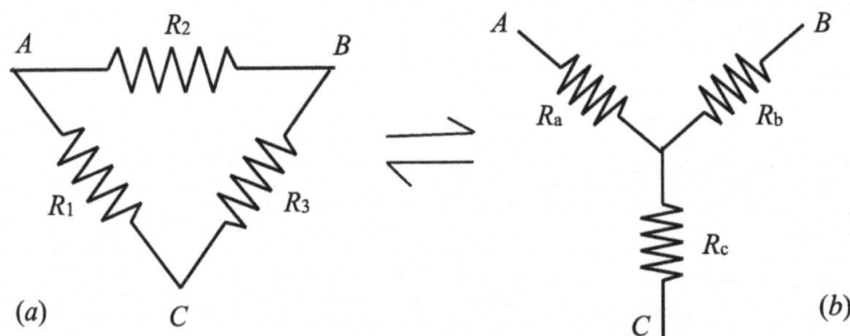

Fig. 2.5 To help obtain inter-conversion between Del and Y configurations of resistors in electrical circuit

$$R_a = \frac{R_1 R_2}{R_1 + R_2 + R_3} \tag{2.15}$$

Equation (2.14) – Eq. (2.11) gives

$$R_b = \frac{R_2 R_3}{R_1 + R_2 + R_3} \tag{2.16}$$

Equation (2.14) – Eq. (2.13) gives

$$R_c = \frac{R_3 R_1}{R_1 + R_2 + R_3} \tag{2.17}$$

Equations (2.15), (2.16) and (2.17) provide one set of conversion, from Del to Y configuration, see Fig. 2.5.

To obtain the other set, we proceed as follows. Taking product of Eqs. (2.15) and (2.16), we get

$$R_a R_b = \frac{R_1 R_2^2 R_3}{(R_1 + R_2 + R_3)^2} \tag{2.18}$$

Taking product of Eqs. (2.16) and (2.17), we get

$$R_b R_c = \frac{R_1 R_2 R_3^2}{(R_1 + R_2 + R_3)^2} \tag{2.19}$$

Taking product of Eqs. (2.17) and (2.15), we get

$$R_c R_a = \frac{R_1^2 R_2 R_3}{(R_1 + R_2 + R_3)^2} \tag{2.20}$$

Summing Eqs. (2.18), (2.19) and (2.20), we get

$$R_aR_b + R_bR_c + R_cR_a = \frac{R_1R_2R_3}{R_1 + R_2 + R_3} \tag{2.21}$$

Now, dividing Eq. (2.21) by Eq. (2.16), we get

$$R_1 = \frac{R_aR_b + R_bR_c + R_cR_a}{R_b} \tag{2.22}$$

Dividing Eq. (2.21) by Eq. (2.17), we get

$$R_2 = \frac{R_aR_b + R_bR_c + R_cR_a}{R_c} \tag{2.23}$$

Dividing Eq. (2.21) by Eq. (2.15), we get

$$R_3 = \frac{R_aR_b + R_bR_c + R_cR_a}{R_a} \tag{2.24}$$

Equations (2.22), (2.23) and (2.24) provide another set of conversion, from Y to Del configuration, see Fig. 2.5.

2.5 Voltage Source and Current Source and Their Inter-conversion

Ideal voltage source can provide same voltage drop V_{S} whatever be the load resistance R_{L} connected across it. See Fig. 2.6. Current through the load resistor changes with change in the resistance of the load resistor, however.

Ideal current source can provide same current I_{S} whatever be the load resistance R_{L} connected in series with it. See Fig. 2.7. Voltage across the load resistor changes with change in the resistance of the load resistor, however.

We now move onto inter-conversion between voltage and current sources. See Fig. 2.8a, b. In Fig. 2.8a, current through the load resistor of resistance R_{L} is $I_{\mathrm{L}} = V_{\mathrm{S}}/(R_{\mathrm{S}} + R_{\mathrm{L}})$ and voltage across it is $V_{\mathrm{L}} = R_{\mathrm{L}}\, I_{\mathrm{L}} = R_{\mathrm{L}}\, V_{\mathrm{S}}/(R_{\mathrm{S}} + R_{\mathrm{L}})$. And in Fig. 2.8b, current through the load resistor of resistance R_{L} given by current division rule is $I_L = \left(\frac{R_s}{R_s+R_L}\right)I_s$

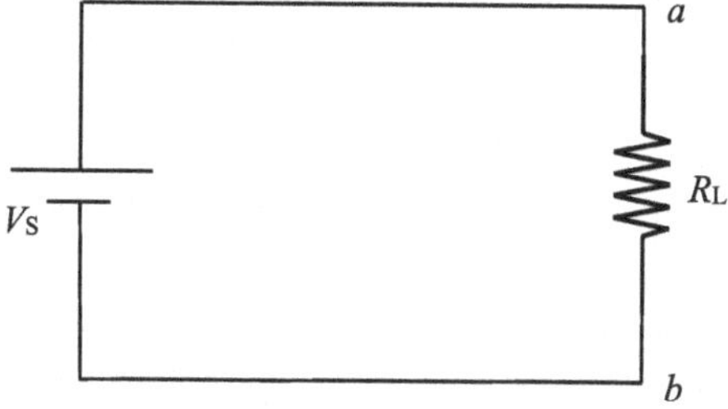

Fig. 2.6 Showing an ideal voltage source V_{S}. Voltage across a and b is $V_{\mathrm{ab}} = V_{\mathrm{S}}$ whatever be the value of resistance of the load resistor R_{L}

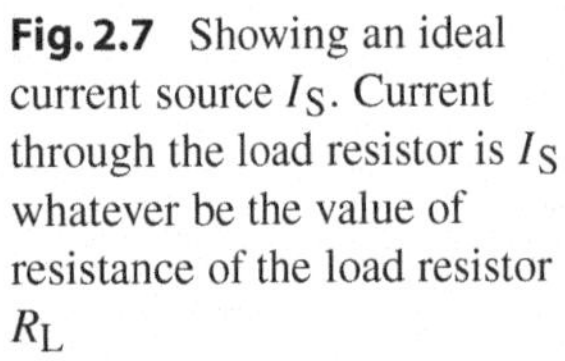

Fig. 2.7 Showing an ideal current source I_S. Current through the load resistor is I_S whatever be the value of resistance of the load resistor R_L

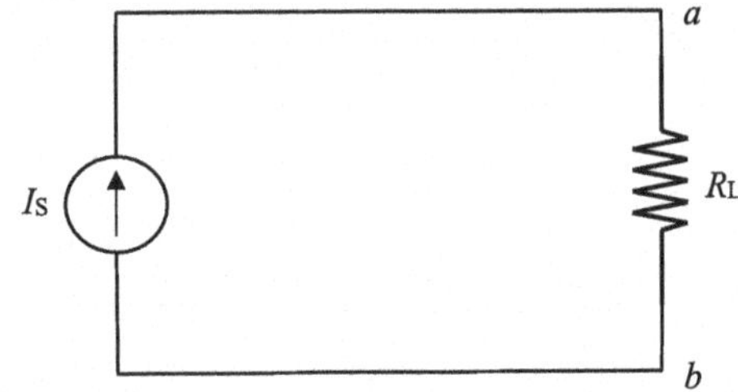

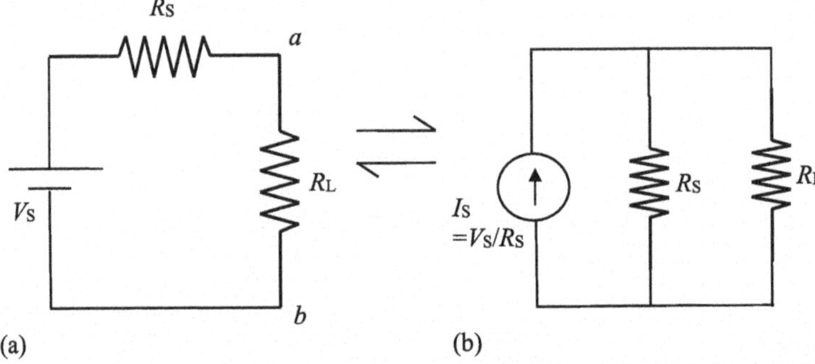

Fig. 2.8 **a** Showing an ideal voltage source V_S. **b** Showing an ideal current source I_S. If $I_S = V_S/R_S$, we can use the two circuits interchangeably. The load resistor R_L will not feel any difference

which for $I_S = V_S/R_S$ becomes $I_L = V_s/(R_s + R_L)$. Voltage across the load is then given by $V_L = R_L I_L = R_L\ V_S/(R_S + R_L)$. Thus current through the load resistors of Fig. 2.8a, b are the same; voltage across the load resistors of Fig. 2.8a, b are the same; if $I_S = V_S/R_S$. Thus we can use the two circuits of Fig. 2.8a, b interchangeably, if $I_S = V_S/R_S$; the load resistor will not feel any difference.

2.6 Superposition Theorem

If we have a circuit containing two or more voltage and/or current sources, current through and/or voltage drop across any resistor of the circuit is equal to algebraic sum of currents or voltage drops produced by each source independently.

To find voltage drop across or current through a given resistor of a circuit due to one of the sources, all other sources have to be removed: voltage sources are replaced by short circuits and current sources are replaced by open circuits.

Let us consider the circuit of Fig. 2.9a. Let us find the current through the resistor of resistance R_1. Due to the battery of emf E_1 alone, current I_{1E1} through R_1 is to be found using the circuit of Fig. 2.9b. This is a simple series circuit. We have

$$I_{1E1} = \frac{E_1}{R_1 + R_2}$$

upwards. Due to the battery of emf E_2 alone, current I_{1E2} through R_1 is to be found using the circuit of Fig. 2.9c. This is also a simple series circuit. We have

$$I_{1E2} = \frac{E_2}{R_1 + R_2}$$

downwards. Due to the current source alone, current $I_{1\text{cs}}$ through R_1 is to be found using the circuit of Fig. 2.9d. We have

$$I_{1cs} = \frac{R_2}{R_1 + R_2} I_s$$

downwards. Thus measured current through R_1 is

$$I_{1E1} \sim (I_{1E2} + I_{1cs}) = \frac{E_1}{R_1 + R_2} \sim \left(\frac{E_2}{R_1 + R_2} + \frac{R_2}{R_1 + R_2} I_s\right)$$

The measured current through R_1 can be upwards or downwards depending on if $I_{1\text{E1}}$ is larger than $I_{1\text{E2}} + I_{1\text{cs}}$ or not respectively.

2.7 Thevenin's Theorem

Figure 2.10a shows any electrical network of resistors and voltage and/or current sources having two output terminals connected to a load resistor. According to Thevenin's theorem, we can replace the circuitry by a series combination of a voltage source V_{Th} and a resistor of resistance R_{Th} as in Fig. 2.10b. If we obtain values of V_{Th} and R_{Th} as prescribed in Thevenin's theorem, load resistor does not feel any difference. Current through and voltage across the load resistor remain unchanged.

Prescriptions to obtain values of V_{Th} and R_{Th} are as follows.

(1) Remove the load resistor R_{L} and leave the terminals a and b open circuited.
(2) Find open circuit voltage across the terminals a and b. This is V_{Th}.
(3) Replace all voltage sources (batteries) by short-circuits.
(4) Replace all current sources by open-circuits.
(5) Find equivalent resistance of the circuit of resistors across terminals a and b. This is R_{Th}.

We now provide an illustration of use of Thevenin's theorem. The circuit to be dealt with is shown in Fig. 2.11a. We 1st remove the load resistor and leave terminals a and b open-circuited as in Fig. 2.11b; voltage drop across R_2 is the open circuit voltage across terminals a and b; this is V_{Th}. Thus

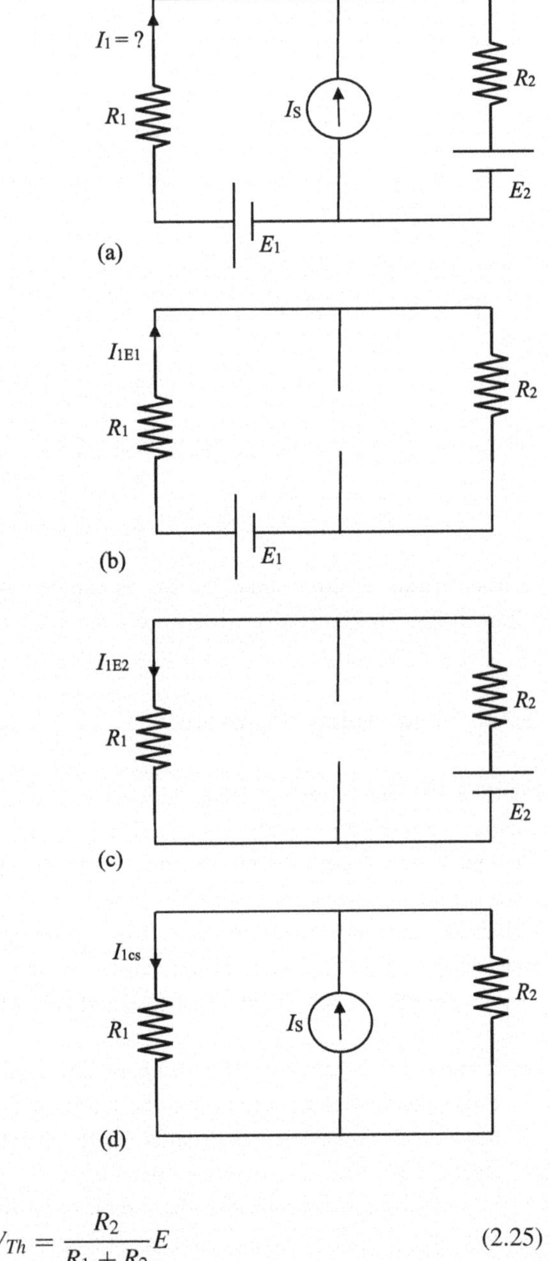

Fig. 2.9 **a** A circuit containing multiple voltage and/or current sources to be analyzed to find current I_1 through R_1. **b** The circuit that determines contribution of voltage source E_1 only to I_1. **c** The circuit that determines contribution of voltage source E_2 only to I_1. **d** The circuit that determines contribution of the current source I_S only to I_1

$$V_{Th} = \frac{R_2}{R_1 + R_2} E \tag{2.25}$$

We now proceed to find out R_{Th}. We resort to Fig. 2.11c in which we have shorted the voltage supply. Equivalent resistance across terminals a and b is R_{Th}. As such

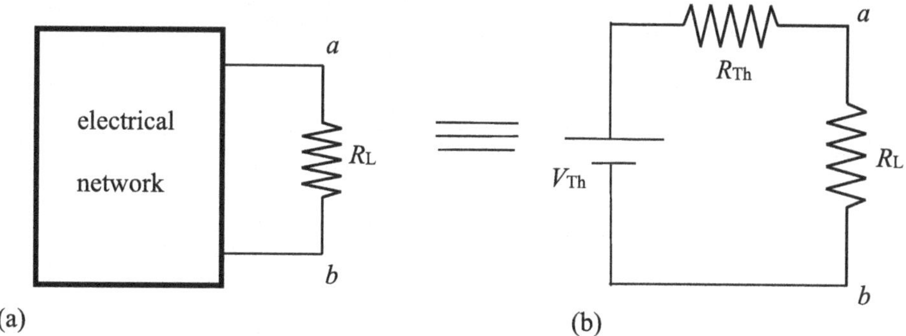

Fig. 2.10 **a** Any electrical network having two output terminals a and b connected to a load resistor of resistance R_L. **b** Thevenin equivalent circuit of the network. With values of E_{Th} and R_{Th} as prescribed in Thevenin's theorem, load resistor does not feel any difference

$$R_{Th} = R_3 + \frac{R_1 R_2}{R_1 + R_2} \tag{2.26}$$

With these values of V_{Th} and R_{Th}, we obtain Fig. 2.11d as Thevenin equivalent circuit of Fig. 2.11a.

See Fig. 2.11d. Voltage across the load resistor is

$$\begin{aligned} V_L &= \frac{R_L}{R_{Th} + R_L} V_{Th} = \frac{R_L}{R_3 + \frac{R_1 R_2}{R_1 + R_2} + R_L} \frac{R_2}{R_1 + R_2} E \\ &= \frac{R_L R_2 E}{R_3(R_1 + R_2) + R_1 R_2 + R_L(R_1 + R_2)} \end{aligned} \tag{2.27}$$

and current through the load resistor is (see Fig. 2.11d)

$$\begin{aligned} I_L &= V_{Th}/(R_{Th} + R_L) = \frac{\frac{R_2}{R_1 + R_2} E}{R_3 + \frac{R_1 R_2}{R_1 + R_2} + R_L} \\ &= \frac{R_2 E}{R_3(R_1 + R_2) + R_1 R_2 + R_L(R_1 + R_2)} \end{aligned} \tag{2.28}$$

We now wish to arrive at Eqs. (2.27) and (2.28) *without* using Thevenin's theorem, but using equivalent resistance laws of series–parallel combination circuit. This will demonstrate validity of Thevenin's theorem.

We start with the same circuit i.e. circuit of Fig. 2.11a drawn again in Fig. 2.12. Here R_2 is in parallel combination with $R_3 + R_L$. As such their equivalent resistance is $\frac{R_2(R_3 + R_L)}{R_2 + (R_3 + R_L)}$ which is in series with R_1. As such total equivalent resistance faced by battery of emf E is

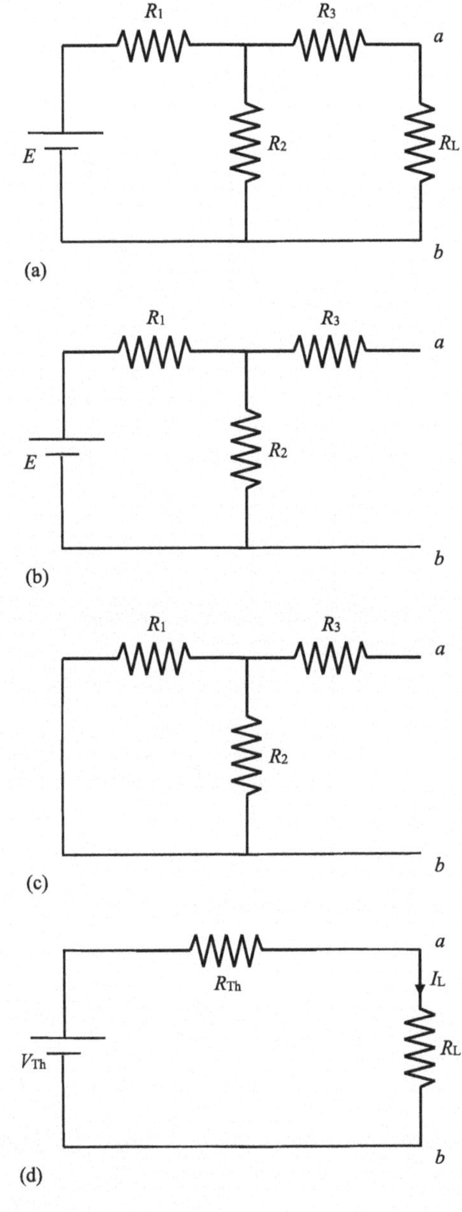

Fig. 2.11 **a** The circuit to be dealt with. **b** The load resistor has been removed; terminals *a* and *b* have been open circuited. **c** Voltage source has been short-circuited. **d** Thevenin equivalent circuit of the circuit in Fig. 2.11a

$$R_1 + \frac{R_2(R_3 + R_L)}{R_2 + (R_3 + R_L)}.$$

Thus the main current is

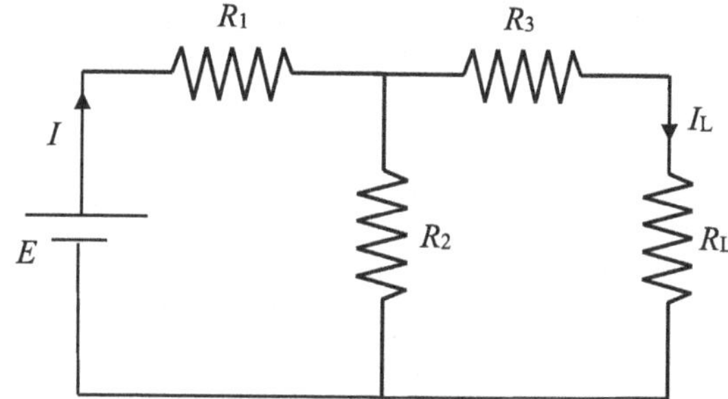

Fig. 2.12 Circuit of Fig. 2.11a redrawn to obtain load current and load voltage drop without using Thevenin's theorem but using equivalent resistance laws of series–parallel combination circuit

$$I = E/\left(R_1 + \frac{R_2(R_3 + R_L)}{R_2 + (R_3 + R_L)}\right)$$

part of which flows through R_L as I_L given by current division rule as

$$\begin{aligned} I_L &= \frac{R_2}{R_2 + R_3 + R_L} I \\ &= \frac{R_2}{R_2 + R_3 + R_L} \frac{E}{R_1 + \frac{R_2(R_3+R_L)}{R_2+(R_3+R_L)}} \\ &= \frac{R_2 E}{R_1(R_2 + R_3 + R_L) + R_2(R_3 + R_L)} \end{aligned} \tag{2.29}$$

As such voltage drop across R_L is

$$V_L = I_L R_L = \frac{R_2 E R_L}{R_1(R_2 + R_3 + R_L) + R_2(R_3 + R_L)} \tag{2.30}$$

Expressions for load current and load voltage drop given by Eqs. (2.29) and (2.30) respectively obtained without using Thevenin's theorem match with those given by Eqs. (2.28) and (2.27) respectively obtained using Thevenin's theorem. The load resistor will not feel any difference. Thus validity of Thevenin's theorem has been demonstrated.

2.8 Norton's Theorem

See Fig. 2.13a, b. Any network of resistors and voltage and/or current sources having two output terminals *a* and *b* can be replaced by a parallel combination of a current source I_S and a resistor of resistance R_N. Load resistor does not feel any difference. Here $R_N = R_{Th}$ the same as that can be obtained using Thevenin's theorem. And I_S is short circuit current at the output terminals *a* and *b*.

We now verify Norton's theorem employing Thevenin's theorem that we have already verified. Using Thevenin's theorem, we can replace the circuitry of Fig. 2.13a by the circuit of Fig. 2.13c. As such current through load resistor is

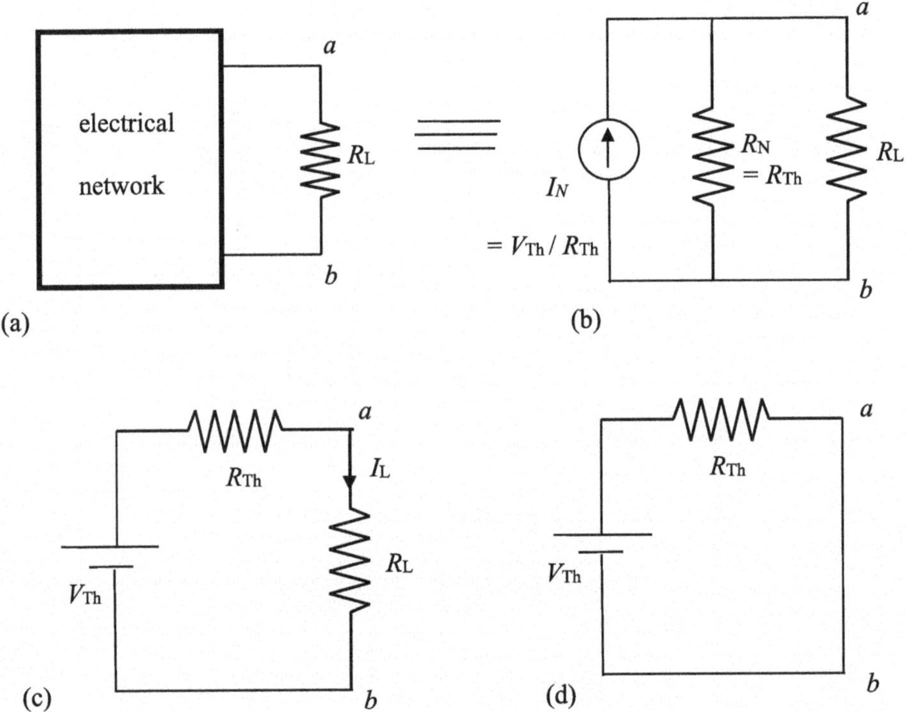

Fig. 2.13 **a** Any electrical network having two output terminals a and b connected to a load resistor of resistance R_{L}. **b** Norton equivalent circuit of the network. With values of $I_N = V_{\mathrm{Th}}/R_{\mathrm{Th}}$ and $R_{\mathrm{N}} = R_{\mathrm{Th}}$ as prescribed in Norton's theorem, load resistor does not feel any difference. **c** Shows Thevenin equivalent circuit of the circuit of **a**. **d** Shows the Thevenin equivalent circuit with the output terminals shorted

$$I_L = \frac{V_{Th}}{R_{Th} + R_L} \tag{2.31}$$

while voltage drop across the load resistor is

$$V_L = \frac{R_L}{R_{Th} + R_L} V_{Th} \tag{2.32}$$

Now see Fig. 2.13d in which we have shorted the two output terminals of the circuit of Fig. 2.13c. The short circuit current is $V_{\mathrm{Th}}/R_{\mathrm{Th}}$. According to Norton's theorem, Norton current in Fig. 2.13b should be equal to this short circuit current which is $I_{\mathrm{S}} = V_{\mathrm{Th}}/R_{\mathrm{Th}}$. With Norton resistance equal to R_{Th}, we now obtain current through and voltage drop across load resistance of the circuit of Fig. 2.13b to show that they are the same as those given by Eqs. (2.31) and (2.32) respectively.

Current division rule applied to the circuit of Fig. 2.13b gives

$$I_L = \frac{R_N}{R_N + R_L} I_N = \frac{R_N}{R_N + R_L} \frac{V_{Th}}{R_{Th}} = \frac{V_{Th}}{R_{Th} + R_L} \quad (2.33)$$

where we have used $R_{\mathrm{Th}} = R_{\mathrm{N}}$. Voltage drop V_{L} across the load in Fig. 2.13b is R_{L} times the load current given by Eq. (2.33). Thus

$$V_L = R_L \frac{V_{Th}}{R_{Th} + R_L} \quad (2.34)$$

Indeed, the voltage drops given by Eq. (2.34) is the same as that given by Eq. (2.32). And indeed the load current given by Eq. (2.33) is the same as that given by Eq. (2.31). Thus we have verified Norton's theorem.

2.9 Maximum Power Transfer Theorem

A general circuit can be replaced by its Thevenin equivalent. Power delivered to the load resistor of resistance R_{L} is given by (Fig. 2.14)

$$P = R_L I_L^2 = R_L \left(\frac{V_{Th}}{R_{Th} + R_L} \right)^2 = \frac{R_L V_{Th}^2}{\left(1 + \frac{R_{Th}}{R_L}\right)^2 R_L^2} = \frac{V_{Th}^2}{\left(1 + \frac{R_{Th}}{R_L}\right)^2 R_L}$$

We can vary the load resistor of resistance R_{L} and can find that the power delivered to R_{L} is maximum for $\frac{dP}{dR_L} = 0$ which gives

$\frac{V_{Th}^2}{R_L} \frac{-2\left(-\frac{R_{Th}}{R_L^2}\right)}{\left(1+\frac{R_{Th}}{R_L}\right)^3} - \frac{V_{Th}^2/R_L^2}{\left(1+\frac{R_{Th}}{R_L}\right)^2} = 0$ or, $\frac{2R_{Th}}{R_L + R_{Th}} = 1$ which gives

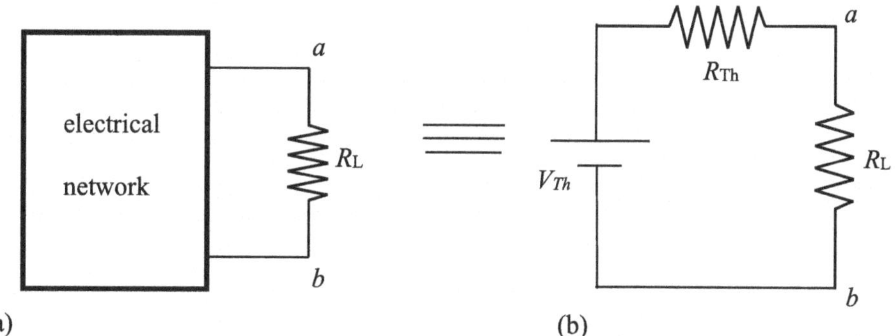

Fig. 2.14 **a** Any electrical network having two output terminals *a* and *b*. **b** Thevenin equivalent of the circuitry of **a**

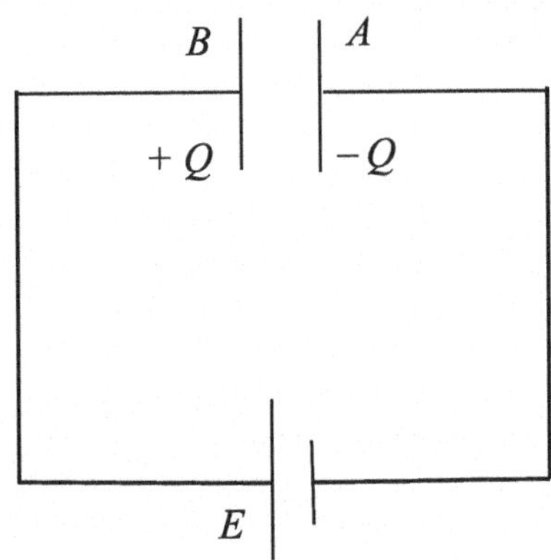

Fig. 2.15 Showing a parallel plate capacitor connected to a DC voltage source or battery of emf E. Here A and B are the two parallel metal plates having air gap or insulator between these

$$R_L = R_{Th} \tag{2.35}$$

It can be shown that $\frac{d^2P}{dR_L^2}$ is negative for $R_L = R_{Th}$. Thus power delivered to the load is maximum if $R_L = R_{Th}$. This is the maximum power transfer theorem.

2.10 Parallel Plate Capacitor in a DC Circuit

Before treating series RC DC circuit (in the next section), it is necessary to understand what happens to a parallel plate capacitor when it is connected to a DC voltage source or battery. See Fig. 2.15. There will be a deficiency of free electrons in plate B and there will be an accumulation of free electrons of same number in plate A. As such there will be an electric field between the two metal plates and hence there will be a potential difference V between the two metal plates. Amount of charge Q on each metal plate is proportional to the potential difference V, i.e. $Q \infty V$ or,

$$Q = CV \tag{2.36}$$

where C is a proportionality constant called capacitance of the capacitor.

Amount of charge Q on each metal plate of the capacitor increases from zero until potential drop V across the capacitor is equal to the emf E of the battery. As such a current flows in the circuit during this time interval. No charge or charged particle physically flows across the gap from one metal plate to the other metal plate of the capacitor. The current flow stops when $V = E$ and then net voltage in the circuit is zero.

2.11 Series *RC* DC Circuit

See Fig. 2.16. If we connect the switch S to terminal A, Kirchhoff's voltage law gives

Fig. 2.16 Showing a switched series RC DC circuit

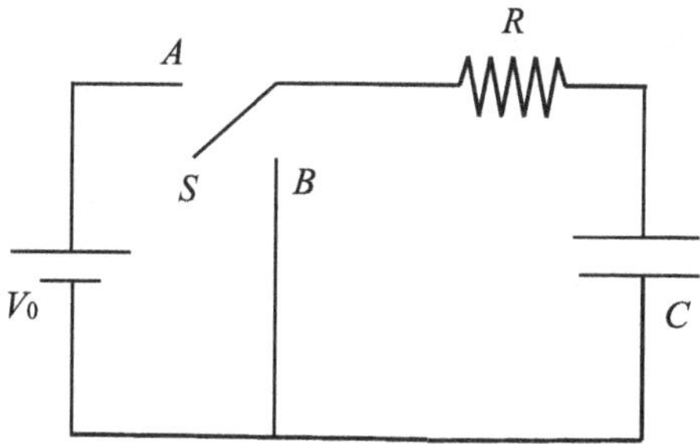

$$V_0 = V_R + V_C \tag{2.37}$$

or,

$$V_0 = RI + Q/C \tag{2.38}$$

Here $V_R = R\,I$ and $V_C = Q/C$ are potential drops across the resistor and the capacitor respectively. I is circuit current and Q is amount of charge on each metal plate of the capacitor. V_0 is emf of the battery.

Differentiation of every term of Eq. (2.38) with respect to time t gives

$$0 = RdI/dt + I/C$$

or,

$$\frac{dI}{dt} + \frac{I}{RC} = 0 \tag{2.39}$$

Solution of this differential equation is

$$I = G\,e^{-\frac{t}{RC}} \tag{2.40}$$

We have the initial condition that at $t = 0$ when we connect the switch S to terminal A (see Fig. 2.16), amount of charge on capacitor plates is zero for which voltage drop across the capacitor is zero. As such Eq. (2.38) gives $V_0 = R\,I_0 + 0$ where I_0 is circuit current at $t = 0$. Thus Eq. (2.40) gives $G = I_0 = V_0/R$. Thus Eq. (2.40) becomes

$$I = \frac{V_0}{R}\,e^{-\frac{t}{RC}} \tag{2.41}$$

Thus circuit current decreases exponentially with time from $I_0 = V_0/R$ to zero. See Fig. 2.17.

During this time interval, voltage drop across the capacitor is given by (see Eqs. 2.37 and 2.38): $V_C = V_0 - V_R = V_0 - R\,I$ or, $V_C = V_0 - V_0\,e^{-\frac{t}{RC}}$ where we have used Eq. (2.41). Thus

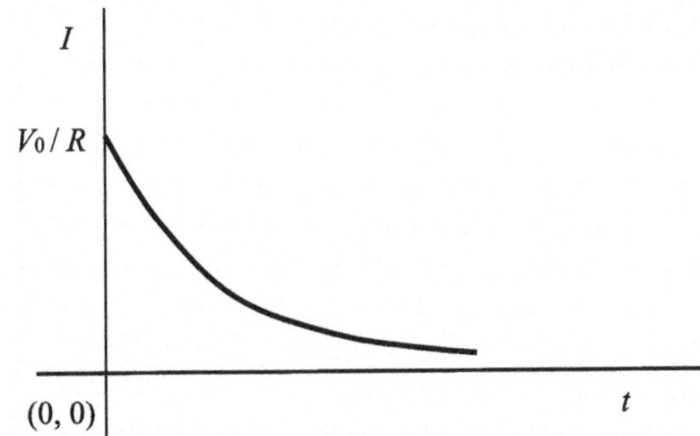

Fig. 2.17 Showing exponential decrease of circuit current as per Eq. (2.41): $I = \frac{V_0}{R} e^{-\frac{t}{RC}}$ from $I_0 = V_0/R$ to zero after connecting the switch S to terminal A in the circuit of Fig. 2.16

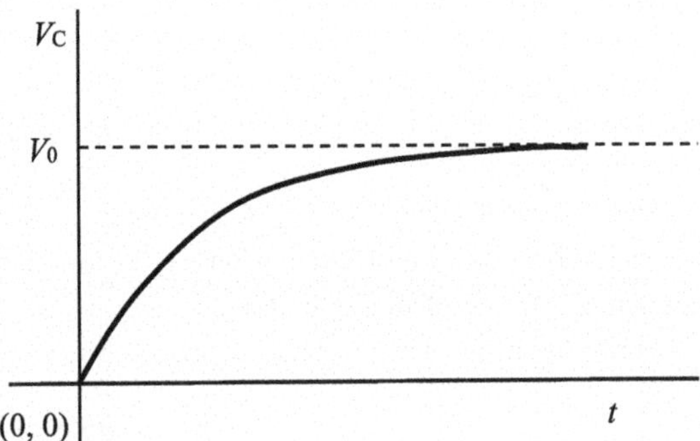

Fig. 2.18 Showing rise of voltage drop across the capacitor as per Eq. (2.42): $V_C = V_0\left(1 - e^{-\frac{t}{RC}}\right)$ from 0 to V_0 after connecting the switch S to terminal A in the circuit of Fig. 2.16

$$V_C = V_0\left(1 - e^{-\frac{t}{RC}}\right) \tag{2.42}$$

We find from Eq. (2.42) that the voltage drop V_C across the capacitor grows from zero to V_0. See Fig. 2.18. It is evident from Eq. (2.42) that the rise of capacitor voltage is faster for smaller value of the product RC. For $R = 0$, it takes zero or no time for V_C to rise from zero to V_0.

After the voltage drop across the capacitor reaches the maximum value V_0, we now turn to the case that the switch S in Fig. 2.16 is connected to terminal B. Kirchhoff's voltage law gives

$$V_R + V_C = 0 \tag{2.43}$$

or,

$$RI + Q/C = 0$$

or,

$$\frac{dI}{dt} + \frac{I}{RC} = 0 \tag{2.44}$$

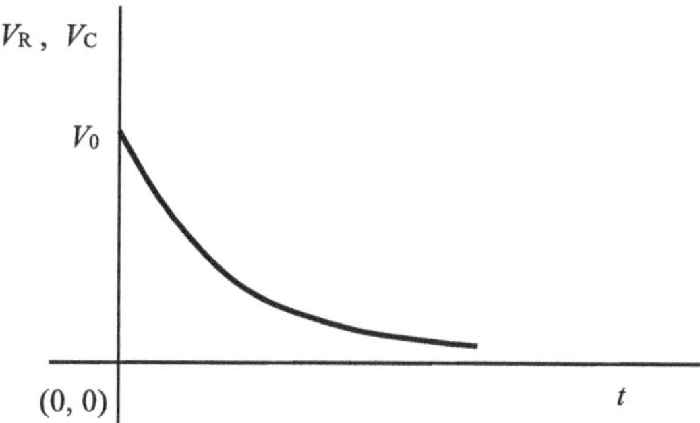

Fig. 2.19 Showing exponential decrease of magnitude of voltage drops across the resistor as well as the capacitor as per Eq. (2.47) or Eq. (2.48): $V_0\, e^{-\frac{t}{RC}}$ from V_0 to zero after connecting the switch S to terminal B in the circuit of Fig. 2.16

where we have taken time derivative. Equation (2.44) is the same as Eq. (2.39) though context is different. Hence solution as before is

$$I = H\, e^{-\frac{t}{RC}} \tag{2.45}$$

We now have the initial condition that at $t = 0$ when we connect the switch S to terminal B of the circuit (see Fig. 2.16), voltage V_0 across the capacitor plates acts as emf and drives current V_0/R through the resistor. Thus Eq. (2.45) gives $H = V_0/R$. Thus Eq. (2.45) gives

$$I = \frac{V_0}{R}\, e^{-\frac{t}{RC}} \tag{2.46}$$

Thus the circuit current exponentially decays with time as in Fig. 2.17. Hence the voltage drop across the resistor of resistance R is given by

$$V_R = RI = V_0\, e^{-\frac{t}{RC}} \tag{2.47}$$

where we have used Eq. (2.46). This is an exponential decrease with time. See Fig. 2.19.

According to Eq. (2.43), magnitude of voltage drop across the capacitor plates is equal to magnitude of voltage drop across the resistor. Thus Eqs. (2.43) and (2.47) gives

$$V_C = V_0\, e^{-\frac{t}{RC}} \tag{2.48}$$

This is also an exponential decrease with time. See Fig. 2.19. Both V_R and V_C decrease exponentially with time and the decrease is slower if the product RC is larger.

AC Circuit Analysis 3

3.1 Introduction to Alternating Current

Magnetic flux through the conducting loop of area A is $\varphi = BA\cos(\theta)$ where θ is angle between area vector which is normal to the surface of the conducting loop and direction of the magnetic field. If the conducting loop rotates with frequency $n = \omega/(2\pi)$, we have $\theta = \omega t$ where t is time. Thus $\varphi = BA\cos(\omega t)$. This variation of magnetic flux with time induces an emf in the conducting loop given by (Fig. 3.1)

$$\varepsilon = -\frac{d\varphi}{dt} = -\frac{d}{dt}(BA\cos(\omega t)) = \omega BA\sin(\omega t) = R_L i \quad (3.1)$$

as per Faraday's law. Thus we get an alternating current i given by

$$i = \frac{\omega BA\sin(\omega t)}{R_L} = I_0\sin(\omega t) \quad (3.2)$$

through the load resistor and an alternating voltage drop

$$v = R_L i = \omega BA\sin(\omega t) = V_0\sin(\omega t) \quad (3.3)$$

across the load resistor.

See Fig. 3.2a. In each period, the voltage drop across the load resistor first rises to maximum value in one direction, falls to zero, polarity of the voltage drop across the load resistor changes; the voltage drop rises to maximum value in the new polarity and again falls to zero.

See Fig. 3.2b. In each period, the current first rises to maximum value in one direction, falls to zero; the direction of current reverses; the current rises to maximum value in the new direction and again falls to zero.

S. Chowdhury, *Introduction to Electronics*, Synthesis Lectures on Engineering, Science, and Technology, https://doi.org/10.1007/978-3-032-03449-6_3

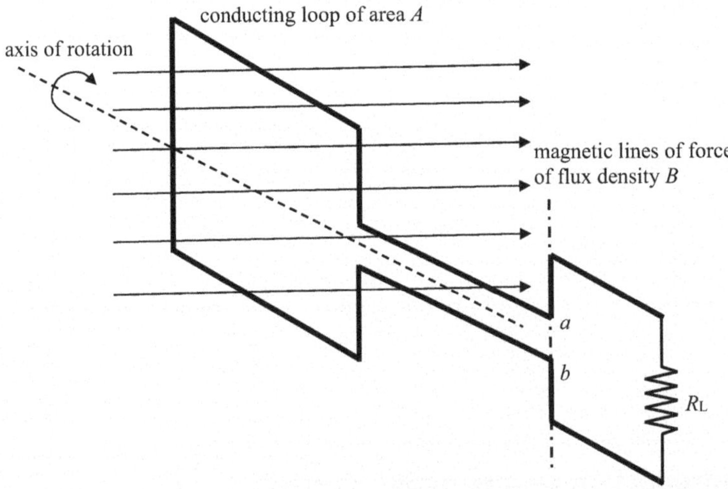

Fig. 3.1 To help introduce production of alternating current. There are arrangements at a and b so that things to the right hand side of dot-dashed vertical line do not rotate with those in the left hand side of the dot-dashed vertical line

Fig. 3.2 **a** Shows variation of alternating voltage drop with time given by $v = V_0 \sin(\omega t)$ and **b** shows variation of alternating current with time given by $i = I_0 \sin(\omega t)$. Here T is time period given by $\omega = 2\pi/T$. Here I_0 and V_0 are called amplitudes of the current and the voltage respectively

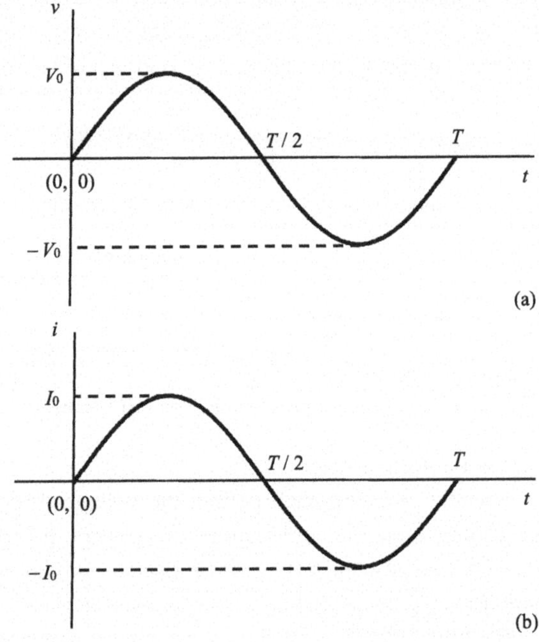

Figure 3.2a, b show one complete cycle of the alternating voltage drop and alternating current respectively. Number of such cycles per unit time is called frequency given by $n = 1/T = \omega/(2\pi)$.

Mean value of the AC current $i = I_0 \sin(\omega t)$ over one time period 0 to T is given by

$$i_{avg} = \frac{1}{T}\int_0^T I_0 \sin\left(\frac{2\pi}{T}t\right) dt = 0$$

and mean value of the AC voltage drop $v = V_0 \sin \omega t$ over one time period 0 to T is given by

$$v_{avg} = \frac{1}{T}\int_0^T V_0 \sin\left(\frac{2\pi}{T}t\right) dt = 0$$

The mean values over one half cycle 0 to $T/2$ are given by

$$i_{avg} = \frac{1}{T/2}\int_0^{T/2} I_0 \sin\left(\frac{2\pi}{T}t\right) dt = 2I_0/\pi$$

and

$$v_{avg} = \frac{1}{T/2}\int_0^{T/2} V_0 \sin\left(\frac{2\pi}{T}t\right) dt = 2V_0/\pi$$

respectively. The so called rms values of the current and the voltage drop are given by

$$i_{rms} = \sqrt{\frac{1}{T}\int_0^T i^2\, dt} = \sqrt{\frac{1}{T}\int_0^T I_0^2 \sin^2\left(\frac{2\pi}{T}t\right) dt} = I_0/\sqrt{2}$$

and

$$v_{rms} = \sqrt{\frac{1}{T}\int_0^T v^2\, dt} = \sqrt{\frac{1}{T}\int_0^T V_0^2 \sin^2\left(\frac{2\pi}{T}t\right) dt} = V_0/\sqrt{2}$$

respectively. The rms values are called effective values because average power delivered to the load resistor in one cycle given by

$$P_{avg} = \frac{1}{T}\int_0^T R_L i^2\, dt = \frac{1}{T}\int_0^T R_L I_0^2 \sin^2\left(\frac{2\pi}{T}t\right) dt = R_L I_0^2/2$$

is the same as that produced if current i were fixed at $I_0/\sqrt{2}$ because $R_L\left(I_0/\sqrt{2}\right)^2 = R_L I_0^2/2$. Same amount of work is done and same amount of Joule heating is produced.

3.2 Resistor in AC Circuit

See Fig. 3.3. We have a source of AC voltage $v_i = V_0 \sin(\omega t)$ connected with a resistor of resistance R. According to Kirchhoff's voltage law, voltage drop v_R across the resistor must be equal to the source voltage v_i at every instant of time so that algebraic sum of the two voltage drops is zero at every instant of time. That is

$$v_R = R\, i_R = v_i = V_0 \sin(\omega t) \tag{3.4}$$

Thus the current i_R through the resistor of resistance R is given by

$$i_R = \frac{V_0}{R} \sin(\omega t) \tag{3.5}$$

Ratio of voltage drop across and current through the resistor is given by

$$\frac{v_R}{i_R} = R \tag{3.6}$$

which is a constant. Thus v_R is proportional to i_R; both go to zero at the same time, both go to maximum value at the same time. See Fig. 3.4.

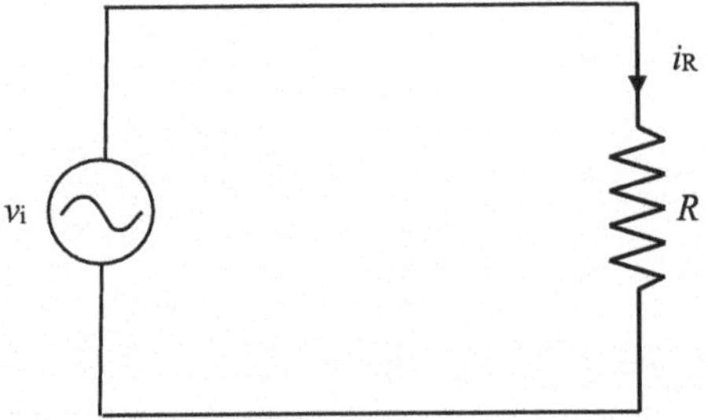

Fig. 3.3 Showing a resistor of resistance R connected with an AC voltage source v_i

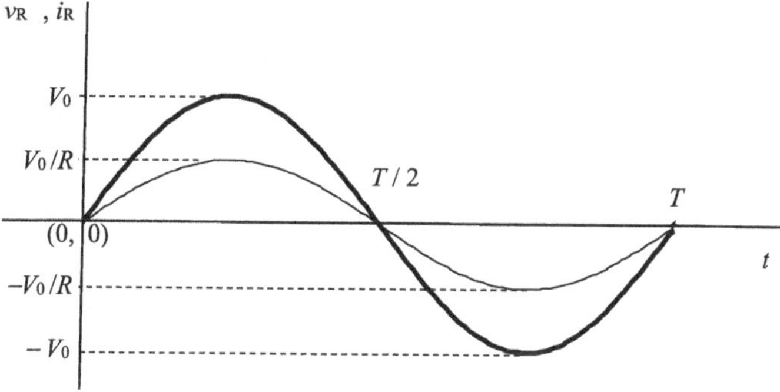

Fig. 3.4 Showing AC voltage drop (thick curve) across the resistor of the circuit of Fig. 3.3 and AC current (thin curve) through the resistor of the circuit of Fig. 3.3 as functions of time. They are in phase; they differ in amplitude only by a factor of R

3.3 Capacitor in AC Circuit

See Fig. 3.5. We have a source of AC voltage $v_i = V_0 \sin(\omega t)$ connected with a capacitor of capacitance C. According to Kirchhoff's voltage law, voltage drop v_C across the capacitor must be equal to the source voltage v_i at every instant of time so that algebraic sum of the two voltage drops is zero at every instant of time. That is

$$v_C = q/C = v_i = V_0 \sin(\omega t) \tag{3.7}$$

Thus amount of charge on each metal plate of the capacitor is

$$q = CV_0 \sin(\omega t) \tag{3.8}$$

and it is a function of time. As such there will be a current i in the connecting wire of the circuit of Fig. 3.5. The current is given by time derivative of the charge q given by

$$i = \omega CV_0 \cos(\omega t) = \omega CV_0 \sin(\omega t + \pi/2) \tag{3.9}$$

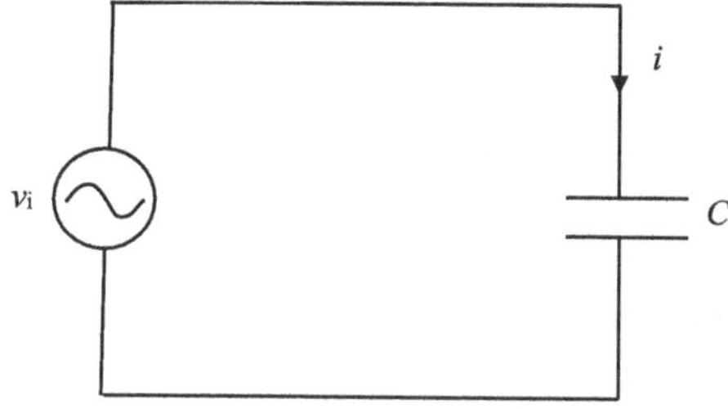

Fig. 3.5 Showing a capacitor of capacitance C connected with an AC voltage source v_i

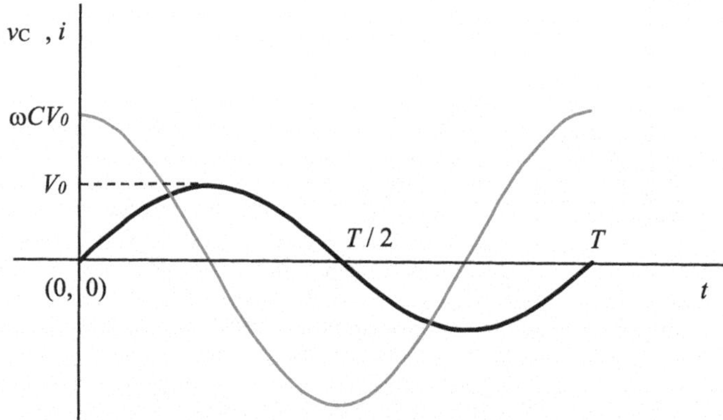

Fig. 3.6 Showing AC voltage drop (thick curve) across the capacitor of the circuit of Fig. 3.5 and AC circuit current (thin curve) of the circuit of Fig. 3.5 as functions of time. They are not in phase; they differ in amplitude also. The phase difference is 90° and the amplitudes differ by a factor of ωC

Voltage drop across the capacitor v_C given by Eq. (3.7) and circuit current i given by Eq. (3.9) differ both in amplitude and phase. The phase difference is 90°. The current *leads* the voltage; voltage *lags* the current.

Ratio of instantaneous values of v_C and i is not constant; it is $\tan(\omega t)/\omega C$. Ohm's law is not valid. Ratio of amplitudes of capacitor voltage and circuit current is $1/\omega C$ and is called *capacitive reactance* $X_C = 1/\omega C = 1/(2\pi f\ C)$ where f is frequency of the AC voltage. X_C is inversely proportional to the frequency f.

Since v_C is not proportional to i; they do not go to zero at the same time, they do not arrive at their maximum values at the same time; see Fig. 3.6. These are expected because when v_C is at its peak value V_0, algebraic sum of it with source voltage is zero; there is no net emf in the circuit and hence no current should be present in the connecting wire. And when v_C is zero, net emf in the circuit is at its peak value V_0; hence current in the connecting wire is at its peak value.

3.4 Inductor in AC Circuit

Before going to analyze response of inductor to AC voltage, we need to narrate response of a single current loop to AC voltage. See Fig. 3.7. Magnetic flux φ through the single loop is proportional to the current i passing through the loop, the constant of proportionality being self-inductance L_1 for single loop. Since the current is time varying, sinusoidal, a back emf is produced given by

$$\varepsilon = -\frac{d\varphi}{dt} = -\frac{d}{dt}(L_1 i) = -L_1\frac{di}{dt} \tag{3.10}$$

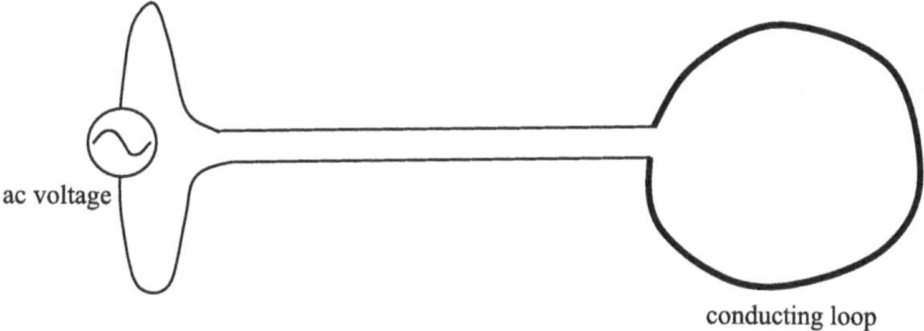

Fig. 3.7 To help introduce response of an inductor to AC voltage

as per Faraday's and Lenz's laws. An inductor is nothing more than a conducting coil of many turns. As such inductance will be larger than that of a single conducting loop.

We now turn to response of an inductor to AC voltage source. See Fig. 3.8. We have a source of AC voltage $v_i = V_0 \sin \omega t$ connected with an inductor of inductance L. According to Kirchhoff's voltage law, voltage drop v_L across the inductor must be equal to the source voltage v_i at every instant of time so that algebraic sum of the two voltage drops is zero at every instant of time. That is

$$v_L = \frac{d\varphi}{dt} = \frac{d}{dt}(Li) = L\frac{di}{dt} = v_i = V_0 \sin \omega t \tag{3.11}$$

Thus circuit current is

$$i = -\frac{V_0}{\omega L} \cos \omega t = \frac{V_0}{\omega L} \sin(\omega t - \pi/2) \tag{3.12}$$

Voltage drop across the inductor v_L given by Eq. (3.11) and circuit current i given by Eq. (3.12) differ both in amplitude and phase. The phase difference is 90°. The current *lags* the voltage; voltage *leads* the current.

Ratio of instantaneous values of v_L and i is not constant; it is $-\ \omega L \tan(\omega t)$. Ohm's law is not valid. Ratio of amplitudes of inductor voltage and circuit current is ωL and it is called *inductive reactance* given by $X_L = \omega L = 2\pi f L$ where f is frequency of the AC voltage. X_L is directly proportional to the frequency f.

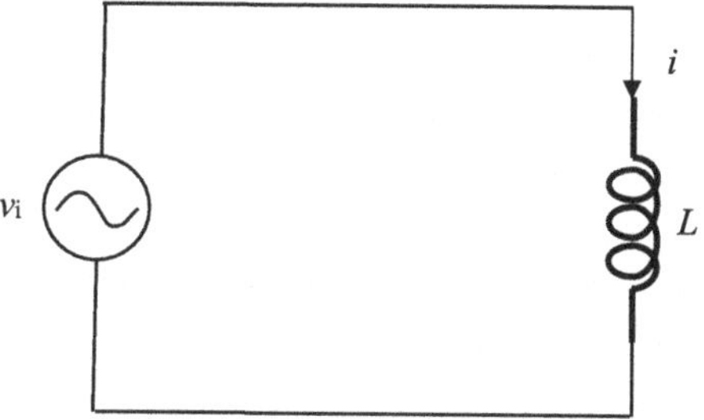

Fig. 3.8 Showing an inductor of inductance L connected with an AC voltage source v_i

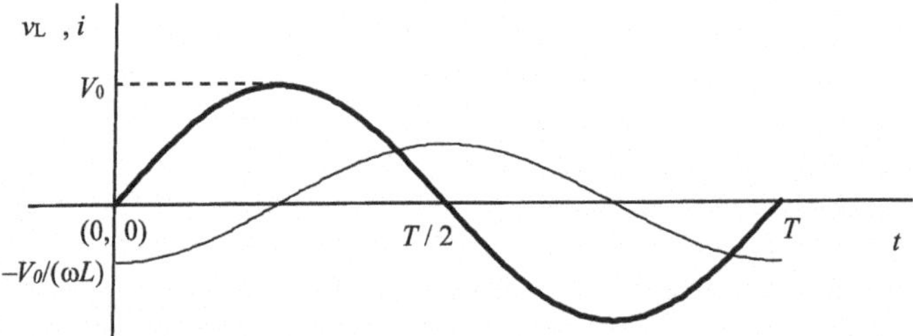

Fig. 3.9 Showing AC voltage drop (thick curve) across the inductor of the circuit of Fig. 3.8 and AC circuit current (thin curve) of the circuit of Fig. 3.8 as functions of time. They are not in phase; they differ in amplitude also. The phase difference is 90° and the amplitudes differ by a factor of ωL

Since v_L is not proportional to i; they do not go to zero at the same time; they do not arrive at their maximum values at the same time; see Fig. 3.9. These are expected because when v_L is at its peak value V_0, algebraic sum of it with source voltage is zero; there is no net emf in the circuit and hence no current should be present in the connecting wire. And when v_L is zero, net emf in the circuit is at its peak value V_0; hence current in the connecting wire is at its peak value.

3.5 Complex Impedance Method of AC Circuit Analysis

We can take input AC voltage as $v_i = V_0 \sin(\omega t)$ or $v_i = V_0 \cos(\omega t)$ or $v_i = V_0 \sin(\omega t + \delta)$ where value of δ is determined by value of v_i at time $t = 0$. Here $v_i = V_0 \sin(\omega t + \delta)$ represents a linear combination of $V_0 \sin(\omega t)$ and $V_0 \cos(\omega t)$. A theoretically valid linear combination of great interest for input AC voltage is $v_i = V_0 \cos(\omega t) + j\, V_0 \sin(\omega t)$ or $v_i = V_0\, e^{j\omega t}$ where $j = \sqrt{(-1)}$.

We now seek response of a resistor to the theoretically valid source of AC voltage $v_i = V_0\, e^{j\omega t}$. See Fig. 3.10. As per Kirchhoff's voltage law, voltage drop across R is $v_R = v_i$ at any instant of time. Thus $R\, i_R = V_0\, e^{j\omega t}$ or $i_R = \frac{V_0}{R}\, e^{j\omega t}$. The ratio of voltage across and current through the resistor is

$$\frac{v_R}{i_R} = \frac{V_0\, e^{j\omega t}}{\frac{V_0}{R}\, e^{j\omega t}} = R \tag{3.13}$$

which is a constant for any given frequency. Ohm's law holds. Voltage across and current through the resistor are in phase.

See Fig. 3.11. We now seek response of a capacitor to the theoretically valid source of AC voltage $v_i = V_0\, e^{j\omega t}$. As per Kirchhoff's voltage law, voltage drop across the capacitor is $v_C = v_i$. Current in the connecting wire of the circuit is

Fig. 3.10 Showing a resistor of resistance R connected with a theoretically valid source of AC voltage $v_i = V_0\, e^{j\omega t}$

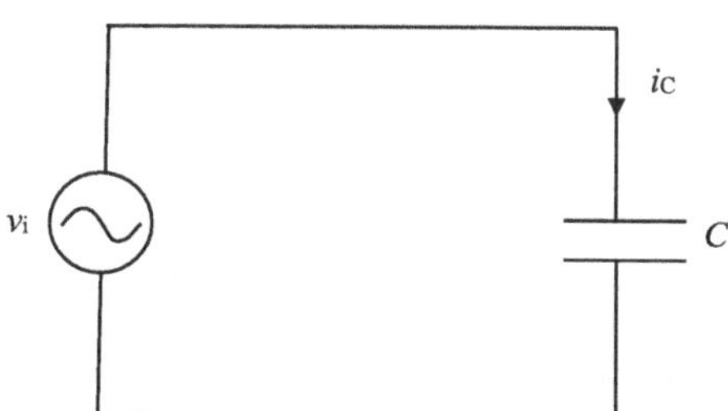

Fig. 3.11 Showing a capacitor of capacitance C connected with a theoretically valid source of AC voltage $v_i = V_0\, e^{j\omega t}$

$$\begin{aligned} i_C &= \frac{dq}{dt} = \frac{d}{dt}(C\, v_C) = C\frac{d}{dt}\left(V_0\, e^{j\omega t}\right) \\ &= j\omega C V_0\, e^{j\omega t} = \omega C V_0\, e^{j(\omega t + \pi/2)} \end{aligned} \tag{3.14}$$

Here q is charge on each plate of the capacitor of capacitance C. Current in the circuit and voltage across the capacitor are not in-phase; they differ in phase by $\pi/2$.

Thus ratio of voltage across the capacitor and current in the circuit is

$$\frac{v_C}{i_C} = \frac{V_0\, e^{j\omega t}}{j\omega C V_0\, e^{j\omega t}} = \frac{1}{j\omega C} \tag{3.15}$$

which is a constant for any certain frequency. Thus Ohm's law holds. But the resistance, now called *impedance*, is a complex quantity $\frac{1}{j\omega C}$. Voltage across the capacitor and current in the circuit are not in phase; they differ in phase by $\pi/2$; see Eq. (3.14). Thus if we use $v_i = V_0\, e^{j\omega t}$ for input AC voltage, we have to use $\frac{1}{j\omega C}$ for impedance of the capacitor.

See Fig. 3.12. We now seek response of an inductor to the theoretically valid source of AC voltage $v_i = V_0\, e^{j\omega t}$. As per Kirchhoff's voltage law, voltage drop across the inductor is $v_L = v_i$. We have $v_L = \frac{d}{dt}(Li_L) = L\frac{di_L}{dt}$. Thus $L\frac{di_L}{dt} = V_0\, e^{j\omega t}$ which gives

$$i_L = \frac{V_0}{j\omega L} e^{j\omega t} = \frac{V_0}{\omega L} e^{j(\omega t - \pi/2)} \tag{3.16}$$

Current in the circuit and voltage across the inductor are not in-phase; they differ in phase by $\pi/2$.

The ratio of voltage across the inductor and current in the circuit is

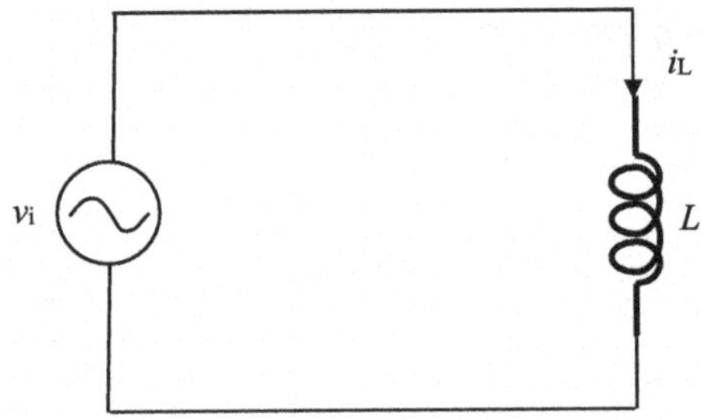

Fig. 3.12 Showing an inductor of inductance L connected with a theoretically valid source of AC voltage $v_i = V_0\, e^{j\omega t}$

$$\frac{v_L}{i_L} = \frac{V_0\, e^{j\omega t}}{\frac{V_0}{j\omega L} e^{j\omega t}} = j\omega L \tag{3.17}$$

which is a constant for any certain frequency. Thus Ohm's law holds. But the resistance, now called *impedance*, is a complex quantity $j\omega L$. Voltage across the inductor and current in the circuit are not in phase; they differ in phase by $\pi/2$; see Eq. (3.16). Thus if we use $v_i = V_0\, e^{j\omega t}$ for input AC voltage, we have to use $j\omega L$ for impedance of the inductor.

If we use $v_i = V_0\, e^{j\omega t}$ as source of input AC voltage, we have to use R, $1/(j\omega C)$ and $j\omega L$ as impedance of resistor, capacitor and inductor respectively. Then we can use Ohm's law for analyzing AC circuits as for resistor circuits with DC voltage. See Fig. 3.13. Equivalent resistance of the parallel combination of L and C is $Z_{LC} = \frac{(j\omega L)\left(\frac{1}{j\omega C}\right)}{j\omega L + \frac{1}{j\omega C}}$ and total impedance faced by the input or source voltage is $Z_{total} = R + \frac{(j\omega L)\left(\frac{1}{j\omega C}\right)}{j\omega L + \frac{1}{j\omega C}}$. Current i_1 is given by $i_1 = v_i/Z_{\text{total}}$. Current i_3 is given by $i_3 = \left(\frac{\frac{1}{j\omega C}}{j\omega L + \frac{1}{j\omega C}}\right) i_1$. Voltage drop across the inductor is $v_L = (j\omega L)\, i_3$. Thus we can analyze AC circuit and get voltage drop across and/or current through any branch of the circuit by calculations as for DC circuits but using $v_i = V_0\, e^{j\omega t}$ as source of input AC voltage; R, $1/(j\omega C)$ and $j\omega L$ as impedance of resistor, capacitor and inductor respectively.

If we wish to get phase difference θ between voltage drop across the resistor and the input voltage e.g., we can calculate the ratio $R\, i_1/v_i$ and give it the form $r\, e^{j\theta}$. Here $i_1 = v_i/Z_{\text{total}}$ and $v_i = V_0\, e^{j\omega t}$.

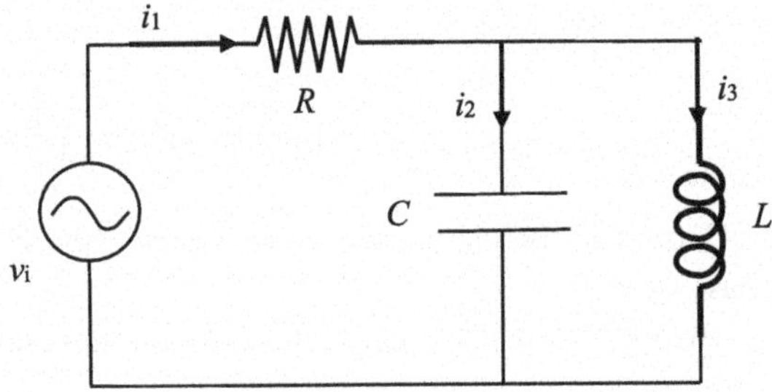

Fig. 3.13 To describe that AC circuits can be analyzed as we do for DC circuits if we use input voltage $v_i = V_0\, e^{j\omega t}$, and ascribe resistances called impedances R, $1/(j\omega C)$ and $j\omega L$ to resistor, capacitor and inductor respectively

The bestowal of $j\omega L$ and $1/(j\omega C)$ as impedance of inductor and capacitor respectively is *not* absolute or fundamental. We can very well take the theoretically valid input AC voltage as $v_i = V_0\, e^{-j\omega t}$. Then impedance that we need to ascribe to inductor and capacitor will be $-j\omega L$ and $1/(-j\omega C)$ respectively. But circuit analyses using either choice yield same information regarding current and voltage.

3.6 Superposition of Oscillatory Functions by Various Methods

Here we show how to obtain result of superposition of two oscillatory functions $y_1 = A \sin(\omega t)$ and $y_2 = B \sin(\omega t + \alpha)$. We do this in 2 different ways; namely by *trigonometric* method and using so called *phasor* method. These methods will be used in the following sections.

We first turn to trigonometric method. Superposition of $y_1 = A \sin(\omega t)$ and $y_2 = B \sin(\omega t + \alpha)$ is given by

$$\begin{aligned} y = y_1 + y_2 &= A\sin(\omega t) + B\sin(\omega t + \alpha) \\ &= A\sin(\omega t) + B(\sin(\omega t)\cos(\alpha) + \cos(\omega t)\sin(\alpha)) \\ &= (A + B\cos(\alpha))\sin(\omega t) + B\sin(\alpha)\cos(\omega t) \end{aligned} \tag{3.18}$$

Let $A + B \cos(\alpha) = P \cos(\beta)$ and $B \sin(\alpha) = P \sin(\beta)$. As such Eq. (3.18) becomes

$$y = P\sin(\omega t)\cos(\beta) + P\cos(\omega t)\sin(\beta) = P\sin(\omega t + \beta) \tag{3.19}$$

Equation (3.19) gives result of superposition of two oscillations $y_1 = A \sin(\omega t)$ and $y_2 = B \sin(\omega t + \alpha)$ where P and β are given by

$$P = \sqrt{(A + B\cos(\alpha))^2 + (B\sin(\alpha))^2} \tag{3.20}$$

and

$$\beta = \tan^{-1}\left(\frac{B\sin(\alpha)}{A + B\cos(\alpha)}\right) \tag{3.21}$$

respectively.

We now turn to *phasor* method. Superposition of $y_1 = A \sin(\omega t)$ and $y_2 = B \sin(\omega t + \alpha)$ is obtained as follows. Two vectors of lengths proportional to A and B are first drawn with angle α between them as in Fig. 3.14. Thereafter, vector P is drawn as in Fig. 3.15. Length of vector P and the angle β that it makes with vector A evidently match with what are given by Eqs. (3.20) and (3.21). This geometric way of obtaining superposition is called phasor method. It often saves laborious expanse of trigonometric calculations.

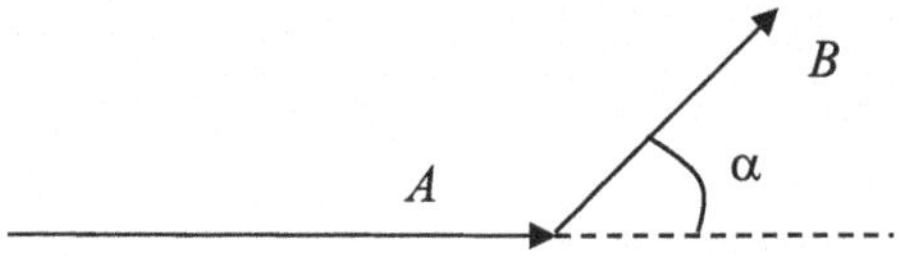

Fig. 3.14 To help illustrate use of phasor method of obtaining superposition of two oscillatory functions

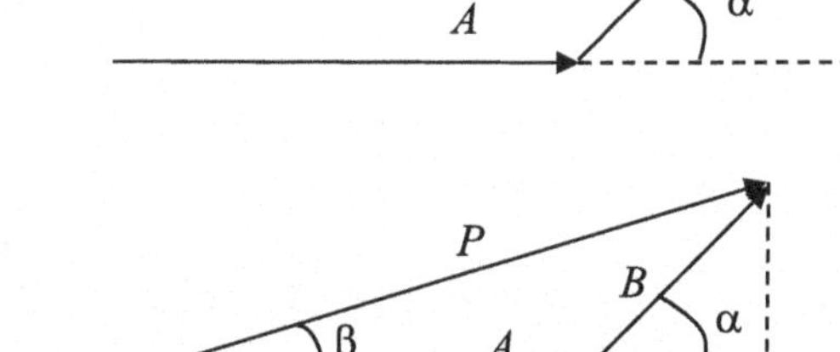

Fig. 3.15 To illustrate use of phasor method of obtaining superposition of two oscillatory functions

3.7 Series *RC* AC Circuit as Filter and as Phase Shifter

See Fig. 3.16. We have a series *RC* AC circuit. We now analyze the circuit using phasor method, trigonometric method and complex impedance method. We will find that the circuit is a high pass filter and positive phase shifter if output voltage is taken across the resistor. And we will find that the circuit is a low pass filter and negative phase shifter if output voltage is taken across the capacitor.

Let the circuit AC current i.e. current through the connecting wire be

$$i = I_0 \sin(\omega t) \tag{3.22}$$

As such voltage drop across the resistor of resistance R is $v_R = R\, i = R\, I_0 \sin(\omega t)$. And voltage drop across the capacitor of capacitance C is $v_C = q/C = \frac{1}{C}\int i\, dt = \frac{1}{C}\int I_0 \sin(\omega t)\, dt$ or,

$$v_C = -\frac{I_0}{\omega C}\cos(\omega t) = \frac{I_0}{\omega C}\sin(\omega t - \pi/2) \tag{3.23}$$

Here q is charge on each metal plate of the capacitor.

According to Kirchhoff's voltage law, sum of v_R and v_C must be equal to v_i, i.e.

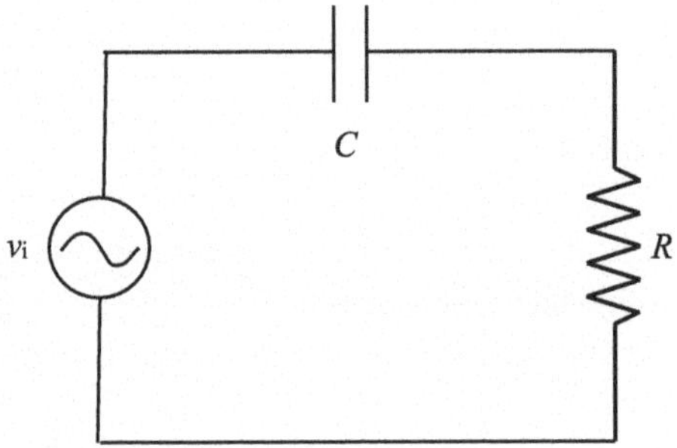

Fig. 3.16 To show that this circuit is a high pass filter and positive phase shifter if output voltage is taken across the resistor. And to show also that this circuit is a low pass filter and negative phase shifter if output voltage is taken across the capacitor

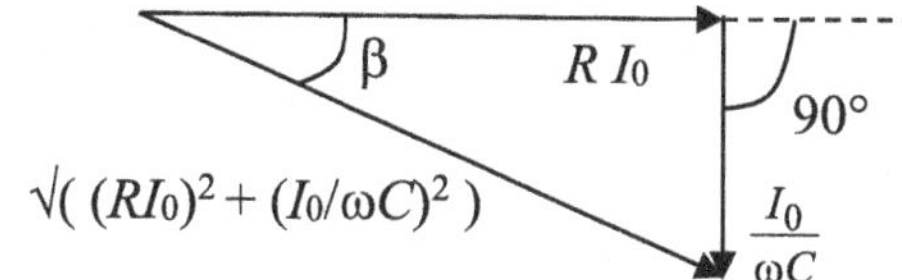

Fig. 3.17 To add the two terms in right hand side of Eq. (3.24) by phasor method

$$v_{\mathrm{i}} = R I_0 \sin(\omega t) + \frac{I_0}{\omega C} \sin(\omega t - \pi/2) \tag{3.24}$$

We now add the two terms in right hand side of Eq. (3.24) by phasor method and subsequently by trigonometric method. Phasor diagram for superposition of the two terms is as in Fig. 3.17. The result is $\surd((RI_0)^2 + (I_0/\omega C)^2)\sin(\omega t - \beta)$ where $\beta = \tan^{-1}\left(\frac{I_0/(\omega C)}{RI_0}\right)$ $= \tan^{-1}\left(\frac{1}{\omega RC}\right)$. Thus $0 < \beta < \pi/2$ for $0 < \omega < \infty$.

Thus we have $v_{\mathrm{i}} = \surd((RI_0)^2 + (I_0/\omega C)^2)\sin(\omega t - \beta)$ or,

$$v_{\mathrm{i}} = \sqrt{\left(R^2 + (1/\omega C)^2\right)} I_0 \sin(\omega t - \beta) \tag{3.25}$$

where

$$\beta = \tan^{-1}\left(\frac{1}{\omega RC}\right) \tag{3.26}$$

We now turn to *trigonometric* method for adding the two terms in right hand side of Eq. (3.24): $R\, I_0 \sin(\omega t) + \frac{I_0}{\omega C}\sin(\omega t - \pi/2)$. We take $P\cos(\beta) = R\, I_0$ and $P\sin(\beta) = I_0/(\omega C)$. As such right hand side of Eq. (3.24) becomes $P\sin(\omega t - \beta)$. Thus Eq. (3.24) becomes

$$v_{\mathrm{i}} = P \sin(\omega t - \beta) \tag{3.27}$$

where $P = \surd((RI_0)^2 + (I_0/\omega C)^2) = \surd(R^2 + (1/\omega C)^2)\, I_0$ and $\beta = \tan^{-1}\left(\frac{1}{\omega RC}\right)$, same as those obtained by phasor method. We gather that $0 < \beta < \pi/2$ for $0 < \omega < \infty$.

We note that if output voltage is taken across the *resistor* we get

$$v_{\mathrm{o}} = v_{\mathrm{R}} = Ri = RI_0 \sin(\omega t) \tag{3.28}$$

which is simple product of R and i, see Eq. (3.22). The input voltage v_{i} is given by Eq. (3.25) or Eq. (3.27). The ratio of their amplitudes is $r = R\, I_0/P = R/\surd(R^2 + (1/\omega C)^2) = 1/\sqrt{(1 + (1/\omega RC)^2)}$ which increases with increasing value of ω. Thus the circuit acts as *high pass filter*. Equations (3.27) and (3.28) also show that output voltage leads the input voltage by $0 < \beta < \pi/2$ for $0 < \omega < \infty$. Thus we have *positive phase shifter*.

We also note that if output voltage is taken across the *capacitor* we get

$$v_{\mathrm{o}} = v_{\mathrm{C}} = \frac{I_0}{\omega C} \sin(\omega t - \pi/2) \tag{3.29}$$

see Eq. (3.23). The input voltage v_i is given by Eq. (3.25) or Eq. (3.27). The ratio of their amplitudes is $r = I_0/(\omega CP) = 1/[\omega\ C\ \sqrt{(R^2 + (1/\omega C)^2)}] = 1/\sqrt{((\omega RC)^2 + 1)}$ which increases with decreasing value of ω. Thus the circuit acts as *low pass filter*. Since $0 < \beta < \pi/2$, Eqs. (3.27) and (3.29) show that output voltage lags the input voltage by $\pi/2 - \beta$ for $0 < \omega < \infty$. Thus we have *negative phase shifter*.

We now analyze the circuit of Fig. 3.16 using complex impedance method. If we theoretically take $v_i = V_0\ e^{j\omega t}$ as the input voltage, we must bestow the resistor and capacitor with impedance R and $1/(j\omega C)$ respectively. We can deal with the problem as for DC circuits. Total impedance faced by the input voltage is $R + 1/(j\omega C)$ which equals ratio of the input voltage v_i and circuit current i. Thus $\frac{v_i}{i} = \frac{V_0\, e^{j\omega t}}{i} = R + 1/(j\omega C)$ which gives

$$i = \frac{V_0\, e^{j\omega t}}{R + 1/(j\omega C)} \text{ or } i = \frac{V_0\, e^{j\omega t}}{R - j/(\omega C)} = \frac{V_0\, e^{j\omega t}}{\sqrt{R^2 + \left(\frac{1}{\omega C}\right)^2}\, e^{-j\beta}}$$

or,

$$i = \frac{V_0\, e^{j(\omega t + \beta)}}{\sqrt{R^2 + \left(\frac{1}{\omega C}\right)^2}} \tag{3.30}$$

Here $\beta = \tan^{-1}(1/(\omega RC))$ and $0 < \beta < \pi/2$.

If output voltage v_o is taken across the *resistor*, $v_o = v_R$ is simple product of R and i. Hence

$$v_o = \frac{RV_0\, e^{j(\omega t + \beta)}}{\sqrt{R^2 + \left(\frac{1}{\omega C}\right)^2}} \tag{3.31}$$

using Eq. (3.30). Ratio of amplitudes of output and input voltages is

$$r = \frac{R}{\sqrt{R^2 + \left(\frac{1}{\omega C}\right)^2}} = \frac{1}{\sqrt{1 + \left(\frac{1}{\omega RC}\right)^2}} \tag{3.32}$$

which increases with increasing value of ω. Thus the circuit acts as *high pass filter*. Since $0 < \beta < \pi/2$, Eq. (3.31) shows that output voltage leads the input voltage $v_i = V_0\ e^{j\omega t}$ by $0 < \beta < \pi/2$ for $0 < \omega < \infty$. Thus we have *positive phase shifter*.

If output voltage v_o is taken across the *capacitor*, $v_o = v_C$ given by

$$\begin{aligned} v_o = v_C = q/C &= \frac{1}{C}\int i\, dt \\ &= \frac{1}{C}\int \frac{V_0\, e^{j(\omega t + \beta)}}{\sqrt{R^2 + \left(\frac{1}{\omega C}\right)^2}} dt \end{aligned}$$

$$= \frac{V_0}{j\omega RC} \frac{e^{j(\omega t+\beta)}}{\sqrt{1+\left(\frac{1}{\omega RC}\right)^2}} = \frac{V_0\, e^{j(\omega t+\beta-\pi/2)}}{\sqrt{1+(\omega RC)^2}} \tag{3.33}$$

using Eq. (3.30). Now ratio of amplitudes of output and input voltage is

$$r = \frac{1}{\sqrt{1+(\omega RC)^2}} \tag{3.34}$$

which increases with decreasing value of ω. Thus the circuit acts as *low pass filter*. Since $0 < \beta < \pi/2$, Eq. (3.33) shows that output voltage lags the input voltage $v_i = V_0\, e^{j\omega t}$ by $\pi/2 - \beta$ for $0 < \omega < \infty$. Thus we have *negative phase shifter*.

3.8 Series *RL* AC Circuit as Filter and as Phase Shifter

See Fig. 3.18. We have a series *RL* AC circuit. We now analyze the circuit using phasor method, trigonometric method and complex impedance method. We will find that the circuit is a low pass filter and negative phase shifter if output voltage is taken across the resistor. And we will find that the circuit is a high pass filter and positive phase shifter if output voltage is taken across the inductor.

Let the circuit AC current i.e. current through the connecting wire be

$$i = I_0 \sin(\omega t) \tag{3.35}$$

As such voltage drop across the resistor of resistance R is $v_R = R\, i = R\, I_0 \sin(\omega t)$. And voltage drop across the inductor of inductance L is

$$v_L = L\frac{di}{dt} = \omega L I_0 \cos(\omega t) = \omega L I_0 \sin(\omega t + \pi/2) \tag{3.36}$$

According to Kirchhoff's voltage law, sum of v_R and v_L must be equal to v_i, i.e.

$$v_i = R I_0 \sin(\omega t) + \omega L I_0 \sin(\omega t + \pi/2) \tag{3.37}$$

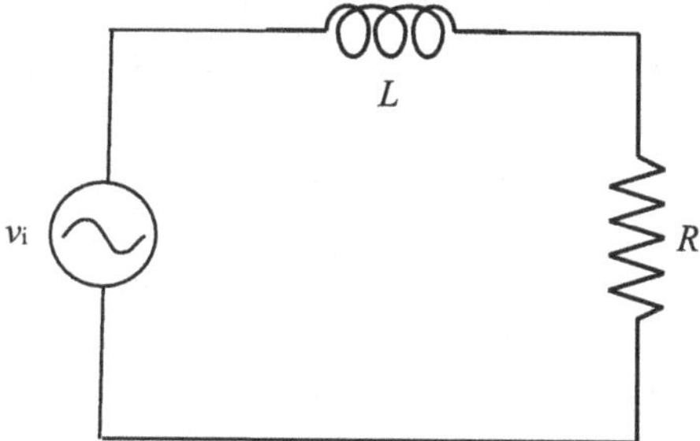

Fig. 3.18 To show that this circuit is a high pass filter and positive phase shifter if output voltage is taken across the inductor. And to show also that this circuit is a low pass filter and negative phase shifter if output voltage is taken across the resistor

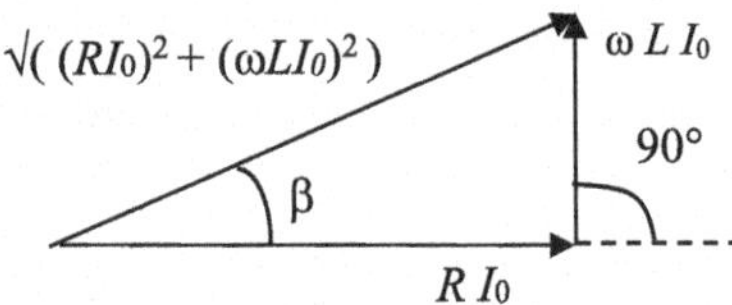

Fig. 3.19 To add the two terms in right hand side of Eq. (3.37) by phasor method

We now add the two terms in right hand side of Eq. (3.37) by phasor method and subsequently by trigonometric method. Phasor diagram for superposition of the two terms is as in Fig. 3.19. The result is $\sqrt{((RI_0)^2 + (\omega L\ I_0)^2)}\ \sin(\omega t + \beta) = \sqrt{(R^2 + (\omega L)^2)}\ I_0\ \sin(\omega t + \beta)$ where $\beta = \tan^{-1}\left(\frac{\omega L I_0}{R I_0}\right) = \tan^{-1}\left(\frac{\omega L}{R}\right)$. Thus $0 < \beta < \pi/2$ for $0 < \omega < \infty$.

Thus we have $v_i = \sqrt{((RI_0)^2 + (\omega LI_0)^2)}\ \sin(\omega t + \beta)$ or,

$$v_i = \sqrt{(R^2 + (\omega L)^2) I_0 \sin(\omega t + \beta)} \tag{3.38}$$

where

$$\beta = \tan^{-1}(\omega L/R) \tag{3.39}$$

We now turn to trigonometric method for adding the two terms in right hand side of Eq. (3.37): $RI_0\ \sin(\omega t) + \omega LI_0 \sin(\omega t + \pi/2) = RI_0\ \sin(\omega t) + \omega LI_0 \cos(\omega t)$. We take $P \cos(\beta) = RI_0$ and $P\ \sin(\beta) = \omega LI_0$. As such right hand side of Eq. (3.37) becomes $P \sin(\omega t + \beta)$. Thus Eq. (3.37) becomes

$$v_{\mathrm{i}} = P \sin(\omega t + \beta) \tag{3.40}$$

where $P = \sqrt{((RI_0)^2 + (\omega LI_0)^2)} = \sqrt{(R^2 + (\omega L)^2)}\ I_0$ and $\beta = \tan^{-1}\left(\frac{\omega L}{R}\right)$, same as those obtained by phasor method. We gather that $0 < \beta < \pi/2$ for $0 < \omega < \infty$.

We note that if output voltage is taken across the *resistor* we get

$$v_{\mathrm{o}} = v_{\mathrm{R}} = Ri = RI_0 \sin(\omega t) \tag{3.41}$$

which is simple product of R and i, see Eq. (3.35). The input voltage v_{i} is given by Eq. (3.38) or Eq. (3.40). The ratio of their amplitudes is $r = RI_0/P = R/\sqrt{(R^2 + (\omega L)^2)} = 1/(1 + (\omega L/R)^2)$ which increases with decreasing value of ω. Thus the circuit acts as *low pass filter*. Equations (3.40) and (3.41) also show that output voltage lags the input voltage by $0 < \beta < \pi/2$ for $0 < \omega < \infty$. Thus we have *negative phase shifter*.

We also note that if output voltage is taken across the *inductor* we get

$$v_{\mathrm{o}} = v_{\mathrm{L}} = L\frac{di}{dt} = \omega LI_0 \sin(\omega t + \pi/2) \tag{3.42}$$

see Eq. (3.35). The input voltage v_{i} is given by Eq. (3.38) or Eq. (3.40). The ratio of their amplitudes is $r = \omega LI_0/P = \omega L/\sqrt{(R^2 + (\omega L)^2)} = 1/\sqrt{(1 + (R/\omega\ L)^2)}$ which increases with increasing value of ω. Thus the circuit acts as *high pass filter*. Since $0 < \beta < \pi/2$, Eqs. (3.40) and (3.42) show that output voltage leads the input voltage by $\pi/2 - \beta$ where $0 < \beta < \pi/2$ for $0 < \omega < \infty$. Thus we have *positive phase shifter*.

We now analyze the circuit of Fig. 3.18 using complex impedance method. If we theoretically take $v_i = V_0\, e^{j\omega t}$ as the input voltage, we must bestow the resistor and inductor with impedance R and $j\omega L$ respectively. We can deal with the problem as for DC circuits. Total impedance faced by the input voltage is $R + j\omega L$ which equals ratio of the input voltage v_i and circuit current i. Thus $\frac{v_i}{i} = \frac{V_0\, e^{j\omega t}}{i} = R + j\omega L$ which gives

$$i = \frac{V_0\, e^{j\omega t}}{R + j\omega L} = \frac{V_0\, e^{j\omega t}}{\sqrt{R^2 + (\omega L)^2}\, e^{j\beta}}$$

or,

$$i = \frac{V_0 e^{j(\omega t - \beta)}}{\sqrt{R^2 + (\omega L)^2}} \tag{3.43}$$

Here $\beta = \tan^{-1}(\omega L/R)$ and $0 < \beta < \pi/2$.

If output voltage v_o is taken across the *resistor*, $v_o = v_R$ is simple product of R and i. Hence

$$v_o = \frac{RV_0\, e^{j(\omega t - \beta)}}{\sqrt{R^2 + (\omega L)^2}} \tag{3.44}$$

see Eq. (3.43). Ratio of amplitudes of output and input voltages is

$$r = \frac{R}{\sqrt{R^2 + (\omega L)^2}} = \frac{1}{\sqrt{1 + (\omega L/R)^2}} \tag{3.45}$$

which increases with decreasing value of ω. Thus the circuit acts as *low pass filter*. Since $0 < \beta < \pi/2$, Eq. (3.44) shows that output voltage lags the input voltage $v_i = V_0\, e^{j\omega t}$ by $0 < \beta < \pi/2$ for $0 < \omega < \infty$. Thus we have *negative phase shifter*.

If output voltage v_0 is taken across the *inductor*, $v_o = v_L$ given by

$$\begin{aligned} v_0 = v_L = L\frac{di}{dt} &= L\frac{d}{dt}\frac{V_0\, e^{j(\omega t - \beta)}}{\sqrt{R^2 + (\omega L)^2}} \\ = j\omega L\frac{V_0\, e^{j(\omega t - \beta)}}{\sqrt{R^2 + (\omega L)^2}} &= \omega L\frac{V_0\, e^{j(\omega t - \beta + \pi/2)}}{\sqrt{R^2 + (\omega L)^2}} \end{aligned} \tag{3.46}$$

using Eq. (3.43). Now ratio of amplitudes of output and input voltages is

$$r = \frac{1}{\sqrt{1 + (R/\omega L)^2}} \tag{3.47}$$

which increases with increasing value of ω. Thus the circuit acts as *high pass filter*. Since $0 < \beta < \pi/2$, Eq. (3.46) shows that output voltage leads the input voltage $v_i = V_0\, e^{j\omega t}$ by $\pi/2 - \beta$ for $0 < \omega < \infty$. Thus we have *positive phase shifter*.

3.9 Series *RLC* AC Circuit as Band-Pass Filter

We now use complex impedance method of analyzing series *RLC* AC circuit. Use of the method is because it offers the opportunity to carry out calculations as in the case of DC circuits except that we take input voltage theoretically as $v_i = V_0\, e^{j\omega t}$ and that we need to bestow resistor, capacitor and inductor with resistance called impedance R, $1/(j\omega C)$ and $j\omega L$ respectively. As such we will be able to carry out analysis of this complex circuit with great ease (Fig. 3.20).

Total impedance faced by the input voltage $v_i = V_0\, e^{j\omega t}$ is $R + j\omega L + 1/(j\omega C)$ which is equal to ratio of the input voltage v_i and circuit current i. Thus

$$\frac{v_i}{i} = \frac{V_0\, e^{j\omega t}}{i} = R + j\omega L + 1/(j\omega C) \tag{3.48}$$

which gives

$$i = \frac{V_0\, e^{j\omega t}}{R + j\omega L + 1/(j\omega C)} = \frac{V_0\, e^{j\omega t}}{R + j\left(\omega L - \frac{1}{\omega C}\right)} = \frac{V_0\, e^{j\omega t}}{\sqrt{R^2 + \left(\omega L - \frac{1}{\omega C}\right)^2}\, e^{j\beta}}$$

or,

$$i = \frac{V_0\, e^{j(\omega t - \beta)}}{\sqrt{R^2 + \left(\omega L - \frac{1}{\omega C}\right)^2}} = I_0\, e^{j(\omega t - \beta)} \tag{3.49}$$

Here

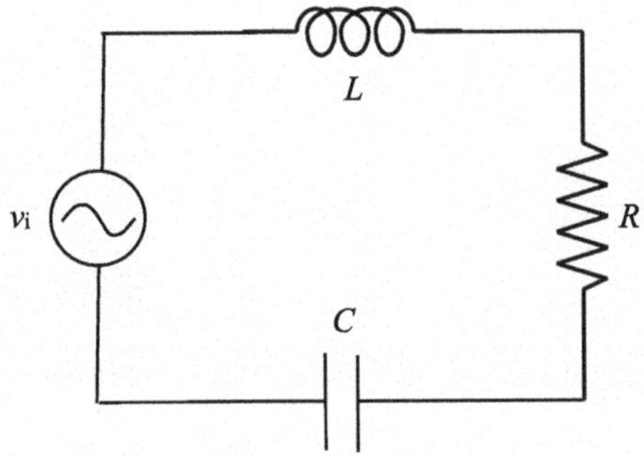

Fig. 3.20 To show that series *RLC* AC circuit acts as band-pass filter

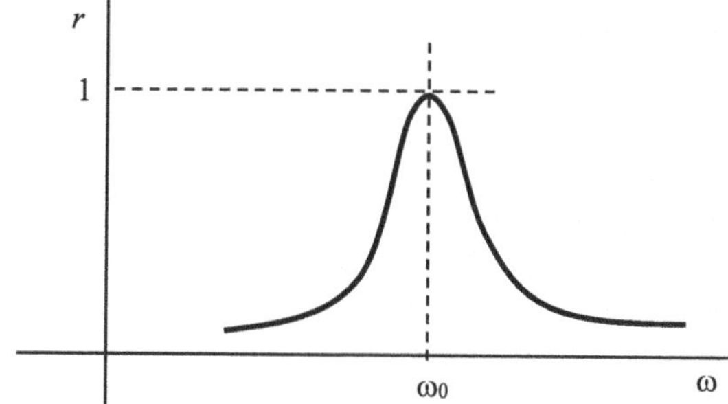

Fig. 3.21 Showing r versus ω plot as per Eq. (3.52): $r = R/\sqrt{R^2 + \left(\omega L - \frac{1}{\omega C}\right)^2}$. The peak is called resonance

$$\beta = \tan^{-1}\left(\frac{\omega L - \frac{1}{\omega C}}{R}\right) \tag{3.50}$$

If output voltage v_o is taken across the resistor, $v_o = v_R$ is simple product of R and i. Hence

$$v_o = v_R = \frac{RV_0 e^{j(\omega t - \beta)}}{\sqrt{R^2 + \left(\omega L - \frac{1}{\omega C}\right)^2}} \tag{3.51}$$

see Eq. (3.49). Ratio of amplitudes of output and input voltages is

$$r = \frac{R}{\sqrt{R^2 + \left(\omega L - \frac{1}{\omega C}\right)^2}} \tag{3.52}$$

which is maximum for $\omega L - \frac{1}{\omega C} = 0$ or $\omega = \frac{1}{\sqrt{LC}}$ called *resonant frequency*. At resonance, total impedance of the circuit is R only as if capacitor and inductor have been removed. r falls to lower values on either side of $\omega = \frac{1}{\sqrt{LC}}$ in r versus ω plot. Thus the circuit acts as a band-pass filter allowing frequencies near resonant frequency to come out in the output voltage across R (Fig. 3.21).

3.10 Series *RC* AC Circuit as Differentiating and Integrating Circuit

Consider the AC circuit of Fig. 3.22. Let the circuit AC current i.e. current through the connecting wire be $i = I_0 \sin(\omega t)$. As such output voltage across the resistor is

$$v_o = v_R = Ri = RI_0 \sin(\omega t).$$

Superposition of this with the AC voltage across the capacitor gives input voltage

$$v_i = R\sqrt{\left(1 + (1/(\omega RC))^2\right)} I_0 \sin(\omega t - \beta)$$

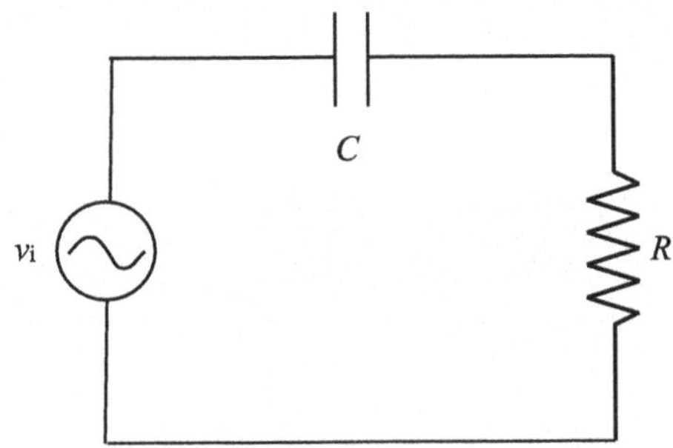

Fig. 3.22 To show that for small values of ωRC, output AC voltage taken across the resistor is time derivative of input AC voltage. And to show that for large values of ωRC, output AC voltage taken across the capacitor is time integral of input AC voltage

where $\beta = \tan^{-1}\left(\frac{1}{\omega RC}\right)$ where $0 < \beta < \pi/2$ for $0 < \omega < \infty$. See Sect. 3.7 for details.

For small values of ωRC, $\beta \approx \pi/2$ and hence $v_i = -(I_0/(\omega C))\cos(\omega t)$ time derivative of which is $(I_0/C)\sin(\omega t) = (RI_0/(RC))\sin(\omega t) = v_o/(RC)$. Thus output AC voltage is RC times time derivative of input AC voltage. Thus the circuit of Fig. 3.22 acts as differentiating circuit for small values of ωRC if output is taken across the resistor.

For large values of ωRC, $\beta \approx 0$ and hence $v_i = RI_0\sin(\omega t)$ time integral of which is

$$\int v_i\,dt = -(RI_0/\omega)\cos(\omega t)$$

Output AC voltage across the capacitor is

$$v_o = v_C = \frac{1}{C}\int i\,dt = \frac{1}{C}\int I_0\sin(\omega t)\,dt = -\frac{RI_0}{\omega RC}\cos(\omega t) = \frac{1}{RC}\int v_i\,dt$$

Thus output AC voltage is $1/(RC)$ times time integral of input AC voltage. Thus the circuit of Fig. 3.22 acts as integrating circuit for large values of ωRC if output is taken across the capacitor.

p–n Junction

4

4.1 Energy Level, Energy Band and Energy Gap

For a Silicon atom, we have 14 electrons in discrete atomic energy levels; see Fig. 4.1a. But in a Silicon crystal, we have allowed energy ranges called energy bands separated by un-allowed or forbidden energy ranges called energy gaps; see Fig. 4.1b. Highest energy band occupied by electrons is called valence band (VB). Lowest empty energy band is called conduction band (CB). The energy difference between lower edge of CB and upper edge of VB is of particular interest and this particular energy gap is denoted by E_g. For Silicon crystal, $E_g = 1.12$ eV.

4.2 Undoped Silicon Crystal

Every Silicon atom in Silicon crystal forms 4 covalent bonds with 4 neighboring Silicon atoms. Each Silicon atom thereby gets 8 outer shared electrons; see Fig. 4.2. At non-zero K temperature, e.g. room temperature 300 K, some of these covalently bonded electrons go from VB to CB by breaking covalent bonds crossing the small energy gap E_g between E_c and E_v. Thus we have free electrons in CB and vacant energy states called holes in VB. Number of free electrons n_e in CB and number of free holes n_h in VB (per unit volume) are equal, given by

$$n_e = n_h = n_i = 2\left(\frac{2\pi k_B T}{h^2}\right)^{3/2} (m_e m_h)^{3/4} e^{-E_g/(2k_B T)} \tag{4.1}$$

where h is Planck's constant, k_B is Boltzmann constant, T is temperature in °K, E_g is energy gap, m_e and m_h are effective mass of electron in CB and hole in VB respectively. For values of parameters appropriate for Silicon crystal, $n_e = n_h = n_i \approx 10^{10}/\text{cm}^3$ at

S. Chowdhury, *Introduction to Electronics*, Synthesis Lectures on Engineering, Science, and Technology, https://doi.org/10.1007/978-3-032-03449-6_4

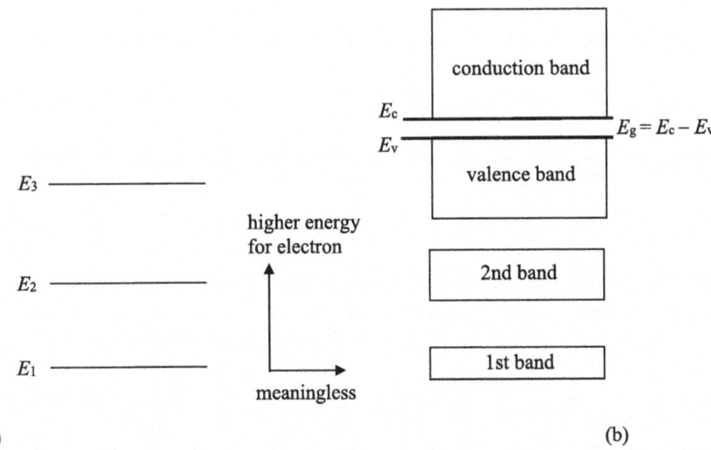

Fig. 4.1 **a** Discrete atomic energy levels $E_1 < E_2 < E_3$ etc. **b** Allowed energy ranges called energy bands separated by un-allowed energy ranges called energy gaps for electrons in atoms in a crystal. E_c is lower edge of conduction band (CB) and E_v is upper edge of valence band (VB). CB and VB are separated by energy gap E_g

300 K. Here i stands for intrinsic or undoped Si crystal. The number n_e and n_h depend on temperature exponentially or strongly.

At non-zero K, if we connect a battery across a piece of Silicon crystal, electrons of both VB and CB start moving towards positive terminal of the battery. As such holes start moving towards negative terminal of the battery. Both electrons in CB and holes in VB contribute to measured current through Silicon crystal. But mobility of hole in VB is smaller than that of electrons in CB. Hence contribution of CB is larger than that of VB. Measured current (in mA scale) is larger for higher temperature because of larger carrier density in both the bands.

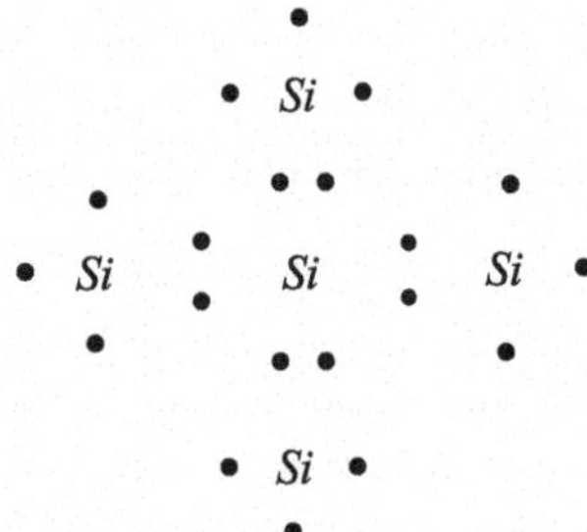

Fig. 4.2 Covalent bonds among neighboring Silicon atoms in Silicon crystal

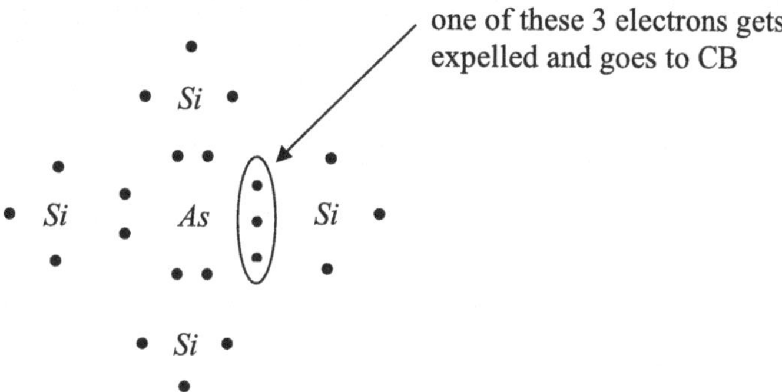

Fig. 4.3 To help explain n-type doping of Silicon crystal by As atoms

4.3 n-Type Si Crystal

If a pentavalent atom e.g. As replaces a Silicon atom of Silicon crystal, we have 9 electrons in place of 8. See Fig. 4.3. One of these 9 electrons gets expelled because of Pauli exclusion principle. The expelled electron comes to CB. Thus number of electrons in CB rises. As atom becomes As^+ ion and is immobile.

Si crystal contains 10^{22} atoms/cm^3. If doping level is 10^{16}/cm^3 (as is usually the case), we have 1 dopant atom replacing 1 out of every 10^6 Si atoms. Number of expelled electrons is 10^{16}/cm^3. Number of electrons in CB becomes $n_e = 10^{10} + 10^{16} \approx 10^{16}$/cm^3. Thus number of electrons in CB rises by a factor of $10^{16}/10^{10} \approx 10^6$. Because of law of mass action which says that product of n_e and n_h remains unaltered even after doping, number of holes in valence band becomes $n_h = (n_i)^2/n_e = (10^{10})^2/10^{16} = 10^4$/cm^3. In summary, CB contains 10^{16} electrons/cm^3 while VB contains 10^4 holes/cm^3 in n-doped Si crystal. Thus electrons in CB are majority carrier while holes in VB are minority carrier.

4.4 p-Type Si Crystal

If a trivalent atom e.g. *B* replaces a Silicon atom of Silicon crystal, we have 7 electrons in place of 8; see Fig. 4.4. We have a vacant site or state near VB edge E_v. Electrons of VB go to this vacant state and we get extra holes in VB. Boron atom becomes B^- ion and is immobile.

Si crystal contains 10^{22} atoms/cm^3. If doping level is 10^{16}/cm^3 (as is usually the case), we have 1 dopant atom replacing 1 out of every 10^6 Si atoms. Number of vacant states is 10^{16}/cm^3. Number of holes in VB becomes $n_h = 10^{10} + 10^{16} \approx 10^{16}$/cm^3. Thus number of holes in VB rises by a factor of $10^{16}/10^{10} \approx 10^6$. Because of law of mass action which says that product of n_e and n_h remains unaltered even after doping, number of electrons

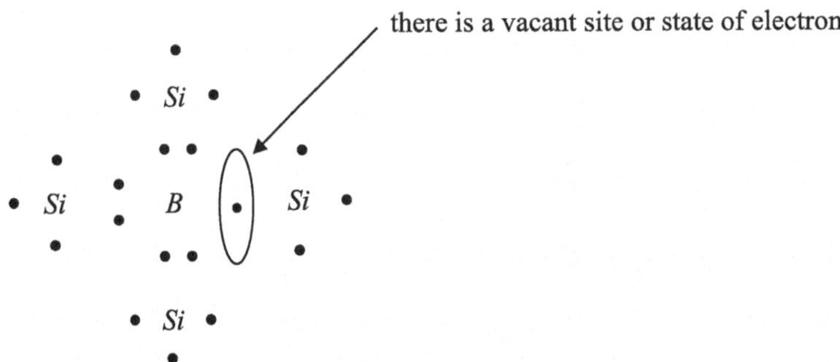

Fig. 4.4 To help explain p-type doping of Silicon crystal by Boron atoms

in CB becomes $n_e = (n_i)^2/n_h = (10^{10})^2/10^{16} = 10^4/\text{cm}^3$. In summary, VB contains 10^{16} holes/cm^3 while CB contains 10^4 electrons/cm^3 in p-doped Si crystal. Thus holes in VB are majority carrier while electrons in CB are minority carrier.

4.5 p–n Junction

If we have a p-Si region and an n-Si region of a Si crystal sharing a common boundary (plane), we have a p–n junction; see Fig. 4.5. In n-Si region, electrons in CB are majority carrier and holes in VB are minority carrier. In p-Si region, holes in VB are majority carrier and electrons in CB are minority carrier. Because of difference in concentration i.e. because of concentration gradient across the junction, diffusion takes place across the p–n junction as indicated by dashed horizontal arrows in Fig. 4.5.

Because of the diffusion, positive donor ions get uncovered in n-type side of the p–n junction while negative acceptor ions get uncovered in p-type side of the p–n junction. See Fig. 4.6. We have a depletion region depleted of free carriers and containing uncovered ions. Electrostatic field builds up inside the depletion region which stops further diffusion at some stage. Because of the electrostatic field, a potential barrier or potential difference builds up across the depletion region. This internal potential barrier is called built-in potential barrier V_{bi}. See Fig. 4.7 and it persists even in absence of applied bias voltage.

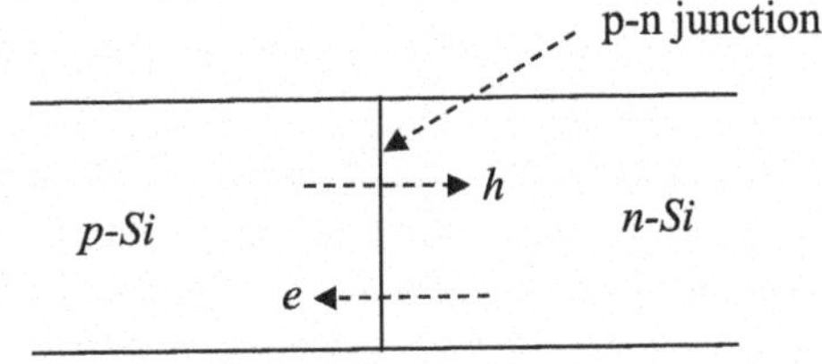

Fig. 4.5 To help introduce p–n junction. Dashed horizontal arrows indicate diffusion of electrons (*e*) and holes (*h*) across the p–n junction

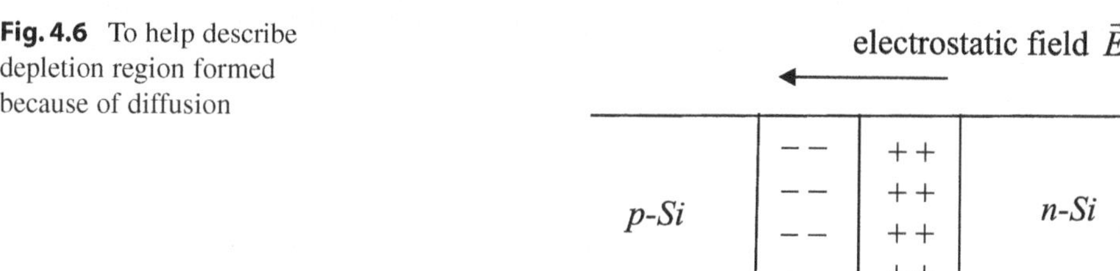

Fig. 4.6 To help describe depletion region formed because of diffusion

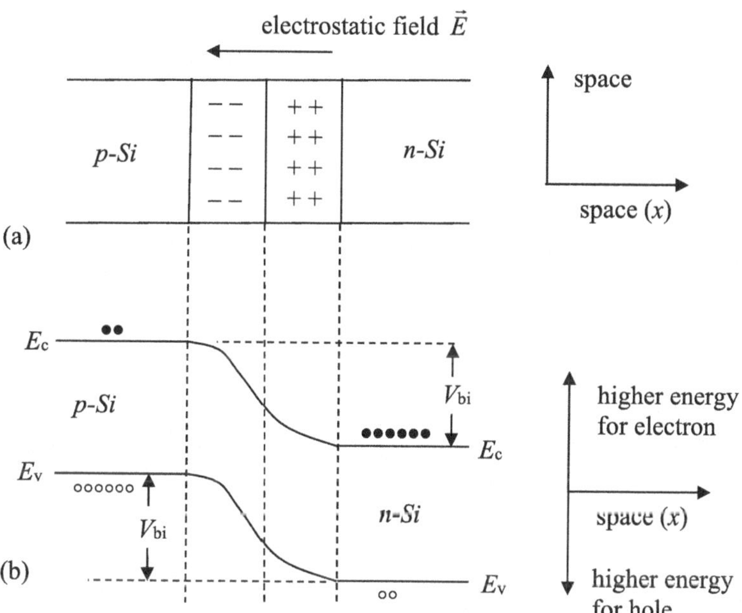

Fig. 4.7 **a** Shows structure and **b** shows band model of p–n junction for *zero* applied bias voltage. **b** Helps in describing built-in potential barrier V_{bi} (≈ 0.7 eV for Silicon crystal) that forms across the depletion region containing uncovered ions indicated in **a**. Here • indicates electron and ○ indicates hole to depict majority and minority carriers. It is indicated that electrons in CB are majority carrier in n-Si and holes in VB are majority carrier in p-Si

At zero applied bias voltage, we have band model of p–n junction as shown in Fig. 4.7b. Majority carrier in CB of n-Si cannot go to CB of p-Si because of the built-in potential barrier. And majority carrier in VB of p-Si cannot go to VB of n-Si because of the built-in potential barrier.

If we apply bias voltage V_{bias} using a battery connecting its negative terminal to n-side and positive terminal to p-side of p–n junction, we call it *forward* bias; it reduces potential barrier height by V_{bias} as indicated in Fig. 4.8b. For $V_{bias} \geq V_{bi}$, we get large current due to both types of majority carriers. The two currents due to the two types of

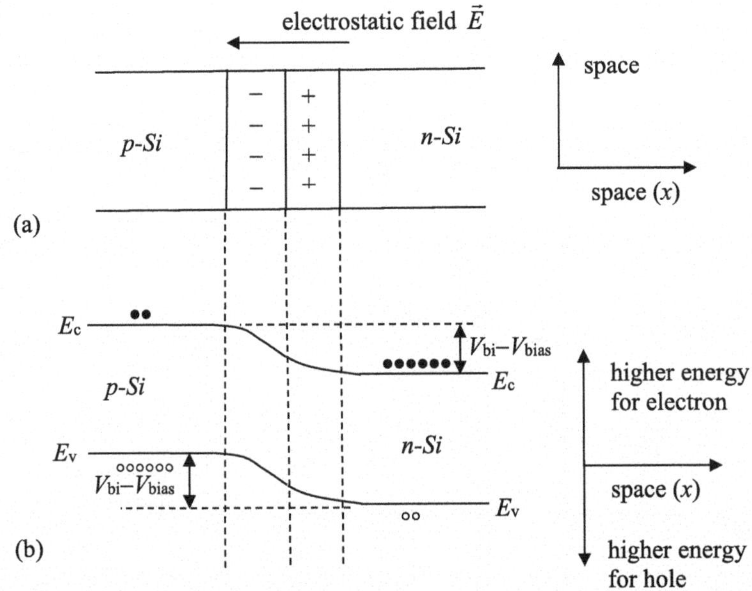

Fig. 4.8 **a** Shows structure and **b** shows band model of p–n junction for applied *forward* bias voltage V_{bias}. As shown in **b**, the potential barrier height reduces to $V_{\text{bi}} - V_{\text{bias}}$ because of the applied forward bias voltage V_{bias}

carriers add to each other and we get measured current in mA scale. Resistance of p–n junction in forward bias is small because conductance, which is proportional to carrier density, is large because of large densities of carriers participating in conduction.

If we apply bias voltage V_{bias} using a battery connecting its positive terminal to n-side and negative terminal to p-side of p–n junction, we call it *reverse* bias. Depletion region widens until potential difference across the depletion region rises to $V_{\text{bi}} + V_{\text{bias}}$. See Fig. 4.9. Majority carriers of both sides are idle; they cannot cross the p–n junction. But minority carriers do cross the junction; but because of their small number, $10^4/\text{cm}^3$, they constitute nA current. Thus conductance is small and hence resistance of p–n junction is high, hundreds of kΩ.

Measured I-V characteristics of p–n junction are as in Fig. 4.10. For forward bias, we have exponential rise of the measured current for $V_{\text{bias}} \geq V_{\text{bi}}$. For reverse bias, we have nA current until a breakdown voltage is reached when reverse current rises sharply for very small rise in reverse voltage. Figure 4.11a shows diode symbol. Arrow-head is in the direction of conventional current when forward biased.

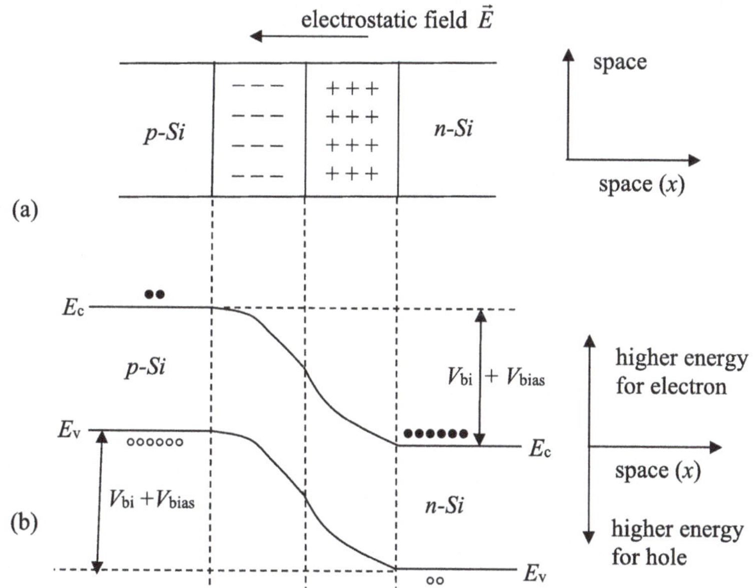

Fig. 4.9 **a** Shows structure and **b** shows band model of p–n junction for applied *reverse* bias voltage V_{bias}. As shown in **a**, depletion region widens and as shown in **b**, the potential barrier height increases to $V_{bi} + V_{bias}$ because of the applied reverse bias voltage V_{bias}

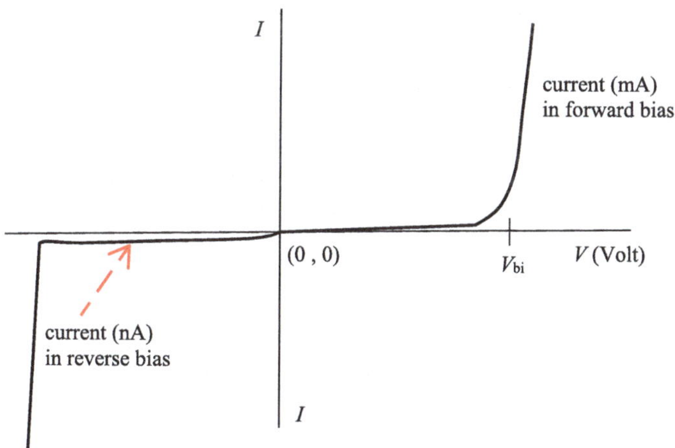

Fig. 4.10 Measured I-V characteristics of p–n junction

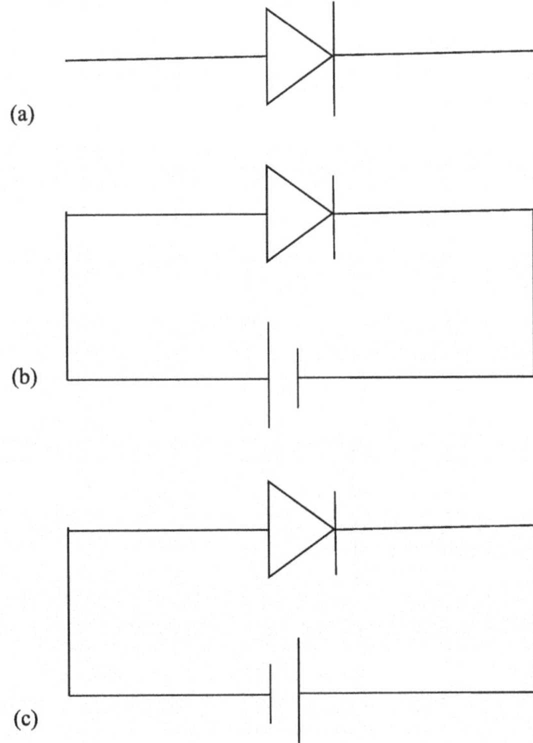

Fig. 4.11 **a** Shows diode symbol, **b** shows diode in forward bias and **c** shows diode in reverse bias

4.6 Half Wave Rectifier

Figure 4.12 shows a half wave rectifier. Input AC voltage is $v_i = V_p \sin(\omega t)$. Output voltage taken across the load resistor R_L is v_o. Output voltage taken across the diode is v_d. These three voltages are shown in Fig. 4.13 as functions of time. We explain the three plots in the following.

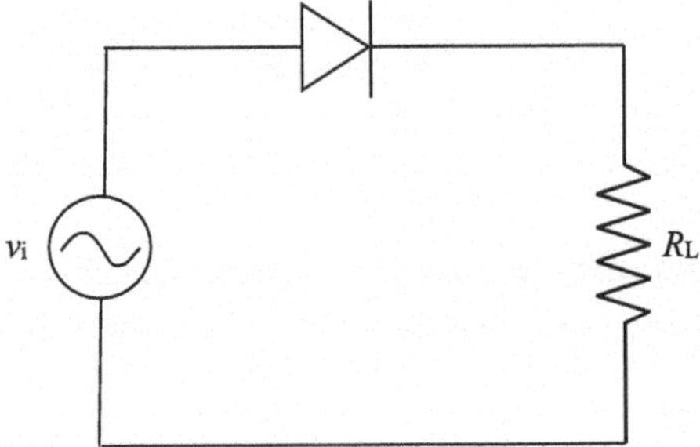

Fig. 4.12 Showing a half wave rectifier. Input AC voltage is $v_i = V_p \sin(\omega t)$. Output voltage taken across the load resistor R_L is v_o. Output voltage taken across the diode is v_d

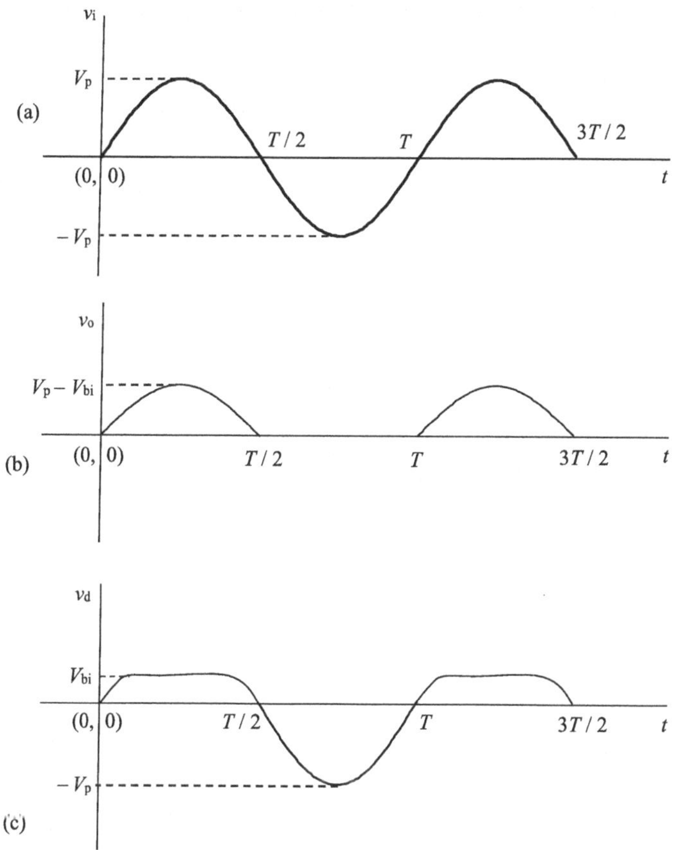

Fig. 4.13 To help explain how a half wave rectifier works. **a** Shows input AC voltage as a function of time. **b** Shows output voltage v_o across the load resistor as a function of time. **c** Shows output voltage v_d across the diode as a function of time

For positive half cycles of the input voltage $v_i = V_p \sin(\omega t)$ which are shown in Fig. 4.13a in the interval $0 < t < T/2$ and $T < t < 3T/2$, the diode is in forward bias and hence maximum voltage drop across the diode is V_{bi}. The rest of the voltage v_i drops across the load resistor R_L. Thus v_o versus time t plot shown in Fig. 4.13b follows the v_i versus t plot except that amplitude of v_o is lower than that of v_i by V_{bi}.

For negative half cycle of the input voltage $v_i = V_p \sin(\omega t)$ which is shown in Fig. 4.13a in the interval $T/2 < t < T$, resistance of the diode is hundreds times larger than R_L because of very small (nA) minority carrier current. Thus almost no voltage drops across the load resistor. Input voltage drops almost entirely across the diode.

We have rectified pulsating DC voltage across the load resistor. See Fig. 4.13b. We now analyze the rectified pulsating DC voltage across the load resistor. The DC voltage level that is present in the v_o versus time plot shown in Fig. 4.13b is given by

$$v_{dc} = \frac{1}{T}\int_0^T v_o\,dt = \frac{1}{T}\left[\int_0^{T/2} v_o\,dt + \int_{T/2}^{T} v_o\,dt\right]$$
$$= \frac{1}{T}\left[\int_0^{T/2} (V_p - V_{bi})\sin(\omega t)\,dt + \int_{T/2}^{T} 0\,dt\right] = \frac{V_p - V_{bi}}{\pi} \quad (4.2)$$

The rms value of non-steady component in v_o versus time plot is given by

$$v_{rms} = \sqrt{\frac{1}{T}\int_0^T (v_o - v_{dc})^2\,dt} = \sqrt{\frac{1}{T}\int_0^T \left(v_o^2 - 2v_{dc}v_o + v_{dc}^2\right)dt}$$
$$= \sqrt{\frac{1}{T}\int_0^T v_o^2\,dt - 2v_{dc}\frac{1}{T}\int_0^T v_o\,dt + v_{dc}^2} = \sqrt{\frac{1}{T}\int_0^T \left(v_o^2\,dt\right) - v_{dc}^2}$$
$$= \sqrt{\frac{1}{T}\int_0^{T/2} \left((V_p - V_{bi})^2 \sin^2(\omega t)\,dt\right) - v_{dc}^2}$$
$$= \sqrt{\frac{1}{T}(V_p - V_{bi})^2 \int_0^{T/2} \left(\sin^2(\omega t)\,dt\right) - v_{dc}^2}$$
$$= \sqrt{\frac{1}{4}(V_p - V_{bi})^2 - \frac{1}{\pi^2}(V_p - V_{bi})^2} \quad (4.3)$$

The so called ripple factor is given by

$$r = \frac{v_{rms}}{v_{dc}} = \frac{\sqrt{\frac{1}{4}(V_p - V_{bi})^2 - \frac{1}{\pi^2}(V_p - V_{bi})^2}}{\frac{1}{\pi}(V_p - V_{bi})} = 1.21 \quad (4.4)$$

4.7 Full Wave Bridge Rectifier

The circuit is shown in Fig. 4.14. Input AC voltage $v_i = V_p \sin(\omega t)$ and output voltage v_o across the load resistor R_L are shown in Fig. 4.15a, b respectively.

For positive half cycles of the input voltage $v_i = V_p \sin(\omega t)$ which are shown in Fig. 4.15a in the interval $0 < t < T/2$ and $T < t < 3T/2$, A is at higher potential than B; see Fig. 4.14; the diodes D_1 and D_2 conduct and current flows downwards through the load

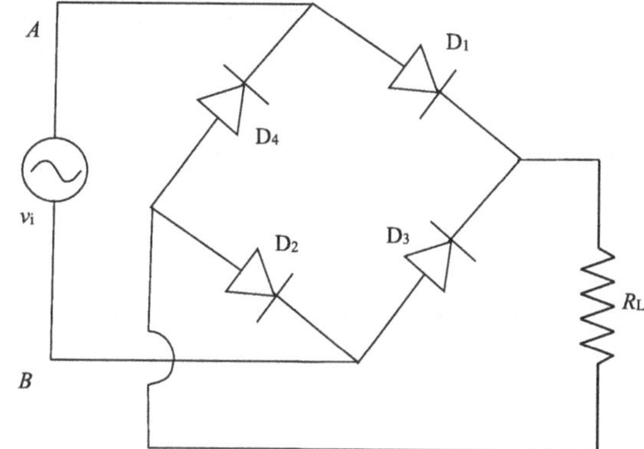

Fig. 4.14 Showing a full wave bridge rectifier. Input AC voltage is $v_i = V_p \sin(\omega t)$. Output voltage taken across the load resistor R_L is v_o

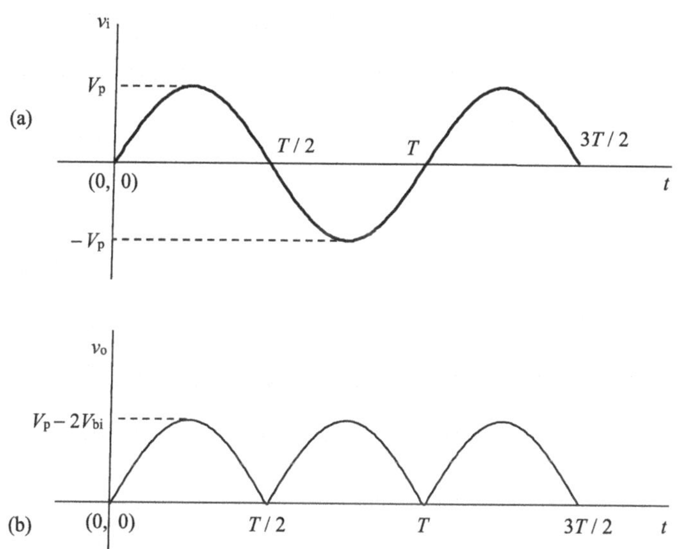

Fig. 4.15 To help explain how a full wave bridge rectifier shown in Fig. 4.14 works. **a** Shows input AC voltage as a function of time. **b** Shows output voltage v_o across the load resistor R_L as a function of time

resistor R_L of Fig. 4.14. Peak value of the voltage across the load resistor is $V_p - 2V_{bi}$ as indicated in Fig. 4.15b. During these time intervals, diodes D_3 and D_4 also conduct but using minority carriers and hence nA current flows upwards through the load resistor of Fig. 4.14. Thus net current (essentially) equals what we get through load resistor via diodes D_1 and D_2.

For negative half cycle of the input voltage $v_i = V_p \sin(\omega t)$ which is shown in Fig. 4.15a in the interval $T/2 < t < T$, B is at higher potential than A; see Fig. 4.14;

the diodes D_3 and D_4 conduct and current flows downwards (again) through the load resistor of Fig. 4.14. Thus we have rectification. Peak value of the voltage across the load resistor is $V_p - 2V_{bi}$ as indicated in Fig. 4.15b. During this time interval, diodes D_1 and D_2 also conduct but using minority carriers and hence nA current flows upwards through the load resistor of Fig. 4.14. Thus net current (essentially) equals what we get through load resistor via diodes D_3 and D_4. Thus we have a complete explanation of full wave rectified output voltage shown in Fig. 4.15b.

We get a pulsating DC voltage as output voltage. See Fig. 4.15b. We now analyze the rectified pulsating DC voltage across the load resistor of the full wave bridge rectifier. The DC voltage level that is present in the output voltage v_o versus time plot shown in Fig. 4.15b is given by

$$v_{dc} = \frac{1}{T/2} \int_0^{T/2} \left(V_p - 2V_{bi}\right) \sin(\omega t)\, dt = \frac{2\left(V_p - 2V_{bi}\right)}{\pi} \tag{4.5}$$

The rms value of non-steady component in output voltage v_o versus time plot is given by

$$v_{rms} = \sqrt{\frac{1}{T/2} \int_0^{T/2} (v_o - v_{dc})^2\, dt} = \sqrt{\frac{\left(V_p - 2V_{bi}\right)^2}{2} - v_{dc}^2} \tag{4.6}$$

The so called ripple factor is given by

$$r = \frac{v_{rms}}{v_{dc}} = \frac{\sqrt{\frac{\left(V_p - 2V_{bi}\right)^2}{2} - v_{dc}^2}}{\frac{2}{\pi}\left(V_p - 2V_{bi}\right)} = 0.48 \tag{4.7}$$

4.8 Smoothing Rectified Output Voltage Using Capacitor Filter

See Fig. 4.16. As the input AC voltage $v_i = V_p \sin(\omega t)$ reaches its peak value of V_p during its 1st positive quarter cycle, the capacitor gets charged to V_p without appreciable delay because total series resistance r of the forward biased diodes D_1 and D_2 is small; the situation is as in Fig. 4.17; time constant rC is small.

As soon as the input AC voltage enters its 2nd quarter, the instantaneous input AC voltage v_i is less than the capacitor voltage V_p. Diodes D_1 and D_2 get reverse biased and assume very large (hundreds of kΩ) resistance and allows only nA current. The capacitor discharges through the much smaller load resistance of R_L; see Fig. 4.18. But since time constant $R_L C$ this time is large, voltage drop across the capacitor reduces very slowly. Thus the voltage across the load resistor remains quite smooth (unchanged).

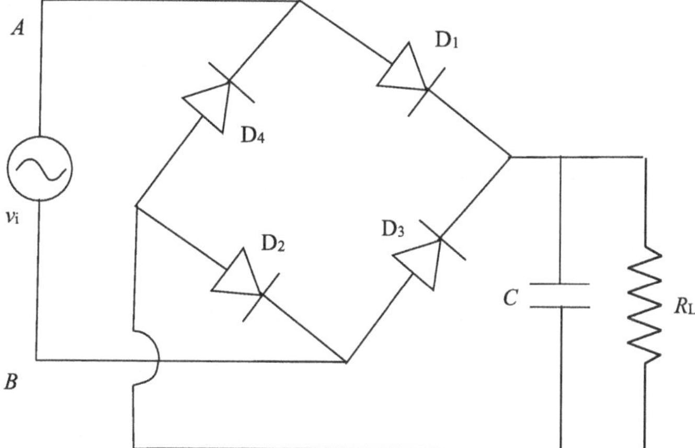

Fig. 4.16 Showing a full wave bridge rectifier with a so called capacitor filter in parallel combination with load resistor. As described in the text, for large value of the product $R_L C$, pulsating output voltage across the load resistor becomes very smooth DC output voltage

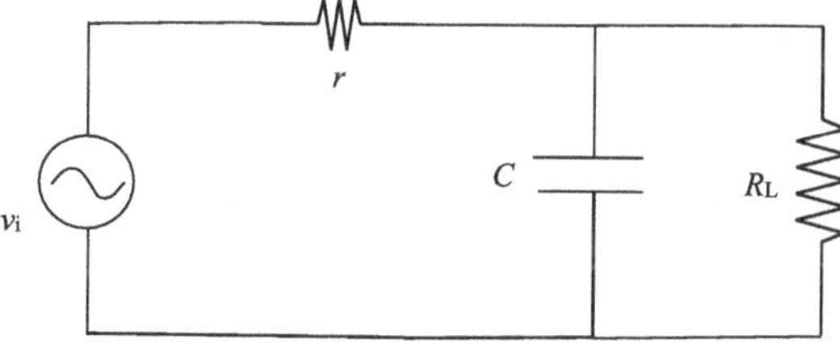

Fig. 4.17 Showing quick charging of the capacitor (to V_P) in series with a resistor of small resistance $r \ll R_L$. See Sect. 2.11 for further details

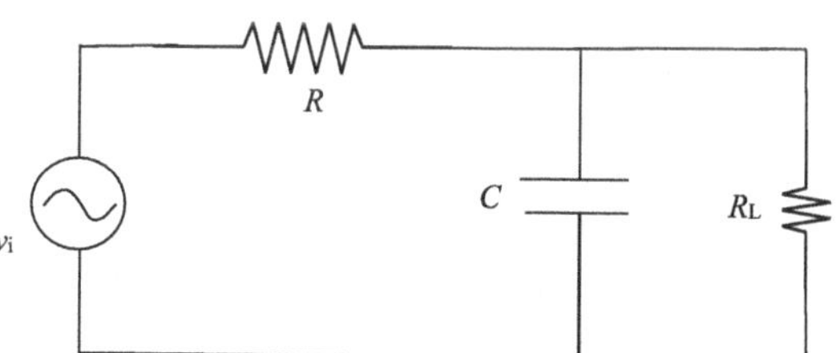

Fig. 4.18 Showing discharge of capacitor voltage through much smaller load resistance $R_L \ll R$. See Sect. 2.11 for further details

The situation remains similar or same for 3rd and 4th quarter cycle of the input voltage except that diodes D_3 and D_4 take up role of diodes D_1 and D_2. During the 3rd quarter cycle, because of small time constant rC, the capacitor charges to V_p quickly with polarity as before and the capacitor voltage remains almost the same during the 4th quarter cycle of the input voltage because of large time constant $R_L C$. Thus we have a very smooth rectified output DC voltage.

4.9 Breakdown of Reverse Biased p–n Junction and Utilizing It in Voltage Regulation

Figure 4.19 reproduces Fig. 4.10; here we have measured I-V characteristics of p–n junction. We pay attention to the reverse biased part. Figure 4.20 reproduces Fig. 4.9 which shows structure and band model of reverse biased p–n junction.

As is evident from the band model in Fig. 4.20b, because of reverse bias, majority carriers are idle while minority carriers do flow (cross the junction) but because of their small number, typically 10^4/cm^3, we have a very small reverse current for $-V_Z < V < 0$.

But as measured I-V data of Fig. 4.19 show, near $V = -V_Z$, there is a sharp rise in reverse current. This is called breakdown of reverse biased p–n junction. This can be because of two possible mechanisms: avalanche breakdown and Zener breakdown.

In large reverse bias, minority carriers get large kinetic energy and collide with atomic electrons and get multiplied in number, called avalanche multiplication and we get a large reverse current. This is avalanche breakdown.

Because of high doping level, width of depletion region of p–n junction can be small and hence average electrostatic field in depletion region can be high. This can cause covalent bond breaking. Thus minority carriers get multiplied in number.

Breakdown voltage depends on doping level. Near $V = -V_Z$, voltage across diode remains almost constant while current through the diode changes hugely from $I_{Z\,\text{knee}}$ to $I_{Z\,\text{max}}$; see Fig. 4.19. We can utilize this feature in what is called voltage *regulation*. See Fig. 4.21. If input voltage V_{IN} varies but $V_{IN} > V_Z$, output voltage V_o remains at V_Z; $V_{IN} - V_Z$ drops across R. reverse current I_Z changes with change in V_{IN}.

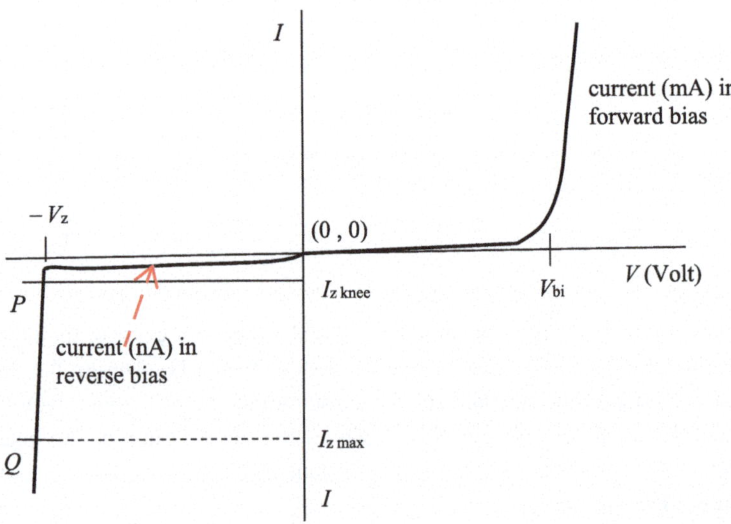

Fig. 4.19 Measured I-V characteristics of p–n junction to analyze the reverse biased part of it

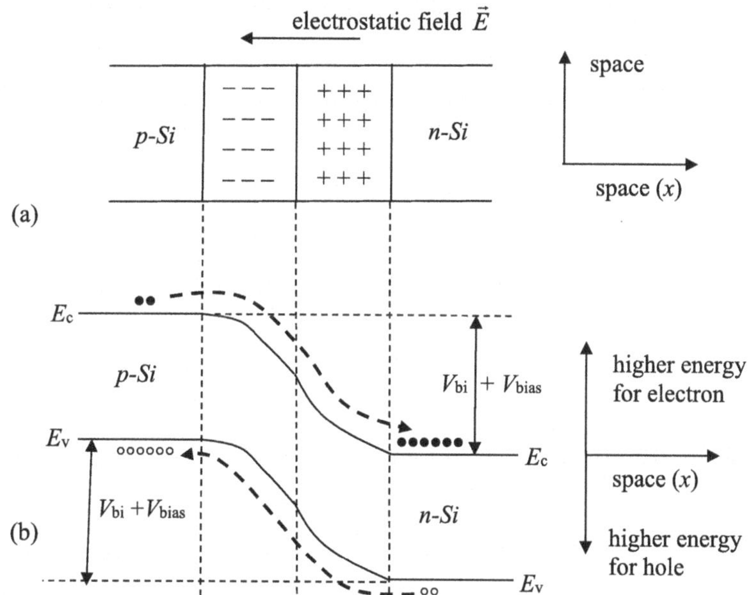

Fig. 4.20 **a** Shows structure and **b** shows band model of p–n junction for applied *reverse* bias voltage V_{bias}. As shown in **a**, depletion region widens and as shown in **b**, the potential barrier height increases to $V_{bi} + V_{bias}$ because of the applied reverse bias voltage V_{bias}. We get flow of minority carriers as indicated by thick dashed arrows

Fig. 4.21 To help explain voltage regulation using p–n junction in breakdown region

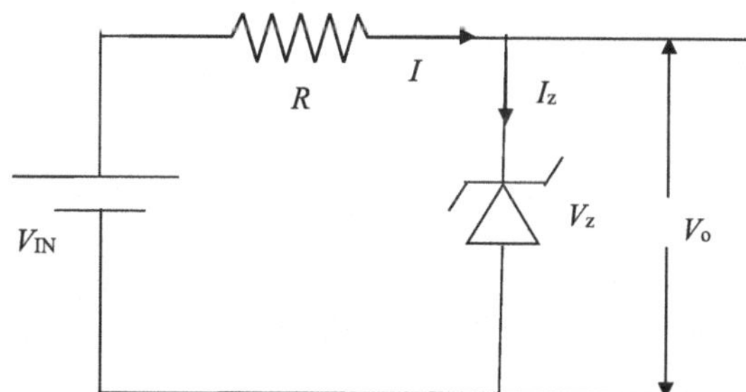

Limitation on variation of V_{IN} is set by the interval $I_{z\,knee} < I < I_{z\,max}$ for Zener diode. If $I_{z\,knee} = 4$ mA, $I_{z\,max} = 40$ mA, $- V_z = 10$ V, $R = 1$ kΩ, minimum allowed value of V_{IN} is $V_z + R\,I_{z\,knee} = 10 + (1\text{ k}\Omega)\,(4\text{ mA}) = 14$ V and maximum allowed value of V_{IN} is $V_z + R\,I_{z\,max} = 10 + (1\text{ k}\Omega)\,(40\text{ mA}) = 50$ V. If V_{IN} varies in the interval 14–50 V, output voltage remains at 10 V. The rest $V_{IN} \sim V_z$ drops across R. Voltage regulation will continue as long as (V, I) is somewhere between P and Q, see Fig. 4.19.

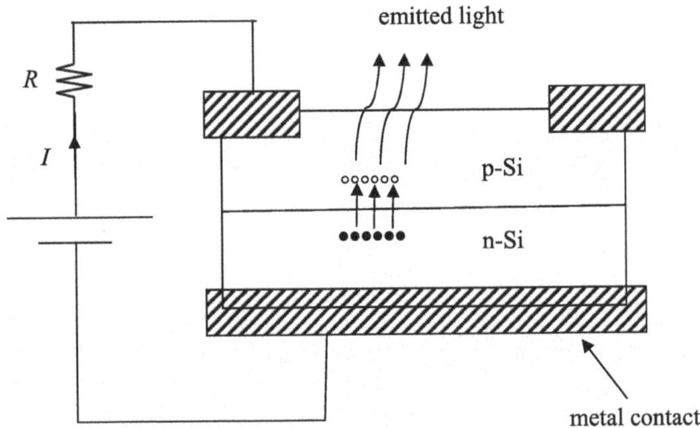

Fig. 4.22 To help describe lighting emitting diode (LED) and explain how it works

4.10 Light Emitting Diode

Figure 4.22 shows structure of light emitting diode (LED) and helps explain how it works. If unbiased, majority carriers cannot recombine sufficiently. Electrons of conduction band of n-Si do not find sufficient number of holes in valence band of n-Si to recombine. And holes in valence band of p-Si do not get sufficient number of electrons in conduction band of p-Si to recombine.

If the p–n junction is forward biased with $V_{\text{bias}} \geq V_{\text{bi}}$, majority carrier (electrons) of conduction band crosses the junction, majority carrier (holes) of valence band also crosses the junction. As such free electrons above E_c find sufficient number of free holes below E_v (on same side of p–n junction) and recombination takes place emitting energy E_g for each recombination as heat or light i.e. photon of energy $E_g \approx h\nu$ (Fig. 4.23).

A large surface area of LED is kept exposed through which light is emitted. The process is called electro-luminescence. Silicon and Germanium are not good for LED. They are heat producing rather than light emitting. Good ones for LED are GaAs, GaP, GaAsP etc.

4.11 Photodiode

Photodiode is p–n junction operated in reverse bias. It has a transparent window that allows light to strike the p–n junction. See Fig. 4.24. In reverse bias, we have a small current flow due to minority carriers. This is called dark current. Because of incident radiation, minority carriers increase in number. This is called generation. These extra generated carriers slide across the p–n junction and constitute larger reverse measured current. For more intense light, we get larger reverse current. See Fig. 4.25.

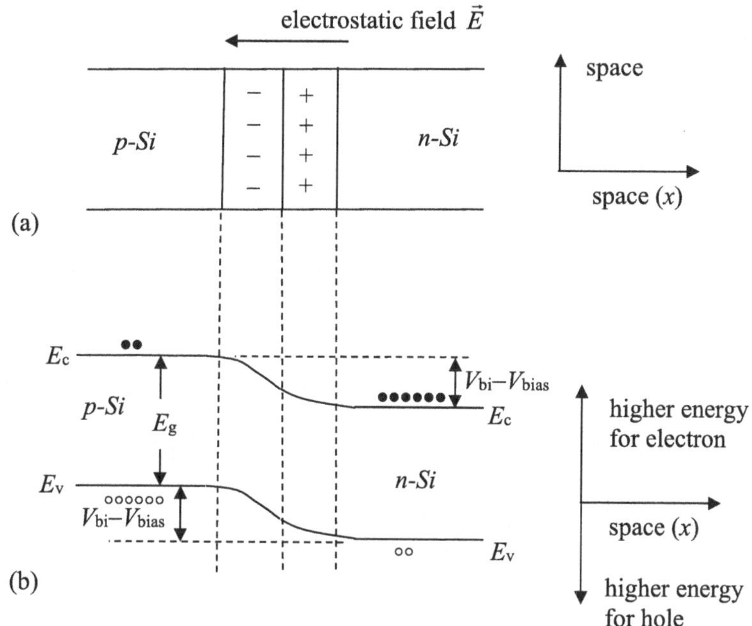

Fig. 4.23 **a** Shows structure and **b** shows band model of p–n junction for applied *forward* bias voltage V_{bias}. As shown in **b**, the potential barrier height reduces to $V_{bi} - V_{bias}$ because of the applied forward bias voltage V_{bias}. For $V_{bias} \geq V_{bi}$, majority carriers cross the p–n junction

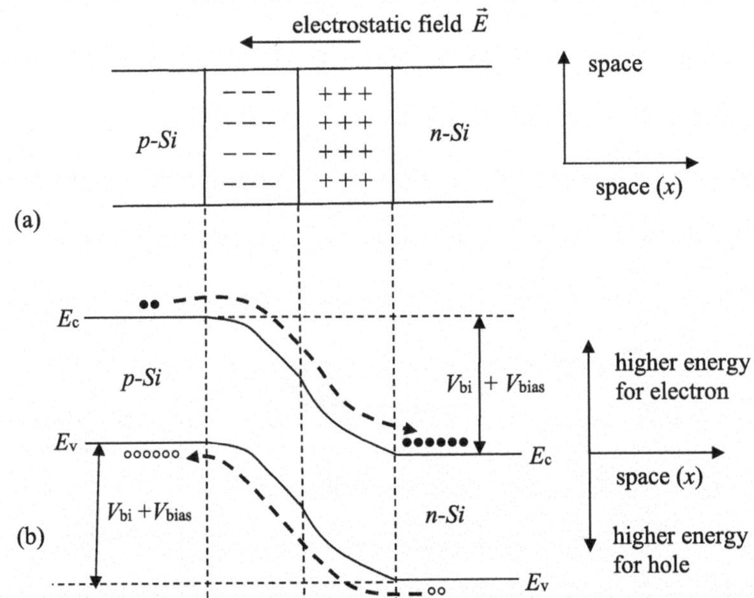

Fig. 4.24 **a** Shows structure and **b** shows band model of p–n junction for applied *reverse* bias voltage V_{bias}. As shown in **a**, depletion region widens and as shown in **b**, the potential barrier height increases to $V_{\text{bi}} + V_{\text{bias}}$ because of the applied reverse bias voltage V_{bias}. We get flow of minority carriers as indicated by thick dashed arrows; the flow is larger for more intense light

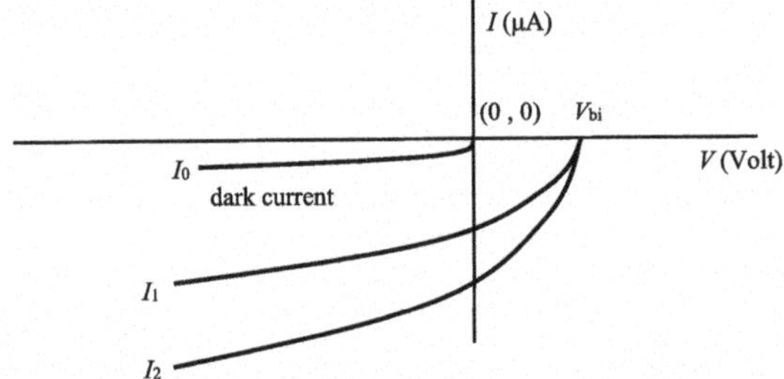

Fig. 4.25 Showing I-V curves for photodiode. Intensity of light $I_2 > I_1 > I_0$ where $I_0 = 0$

Bipolar Junction Transistor (BJT) 5

5.1 Structure of BJT

See Fig. 5.1a; an npn BJT consists of a thin layer of p-type semiconductor crystal sandwiched between two thick layers of n-type semiconductor crystal. See Fig. 5.1b; a pnp BJT consists of a thin layer of n-type semiconductor crystal sandwiched between two thick layers of p-type semiconductor crystal. Annealed metal contacts are formed for so called emitter (E), base (B) and collector (C) terminals as indicated in Fig. 5.1. Emitter and collector are heavily doped but base is lightly doped.

5.2 Band Models of BJT

See Fig. 5.2a; diffusion of charge carriers takes place across each p–n junction. Depletion regions form containing uncovered ions near each p–n junction. As such band model becomes as in Fig. 5.2b for zero applied bias voltage.

But if we apply forward bias V_{BE} to emitter–base junction and reverse bias V_{CB} to base–collector junction as in Fig. 5.3, band model becomes as in Fig. 5.4. Because of applied forward bias V_{BE} to emitter–base junction, potential barrier height across emitter–base junction reduces to $V_{\mathrm{bi}} - V_{\mathrm{BE}}$ and width of the barrier reduces. And because of applied reverse bias V_{CB} to base–collector junction, potential barrier height across base–collector junction rises to $V_{\mathrm{bi}} + V_{\mathrm{CB}}$ and width of the barrier increases.

The circuit configuration shown in Fig. 5.3 is called common base (CB) because base is common to both emitter and collector. Emitter–base junction is forward biased by V_{BE} and base–collector junction is reverse biased by V_{CB}.

S. Chowdhury, *Introduction to Electronics*, Synthesis Lectures on Engineering, Science, and Technology, https://doi.org/10.1007/978-3-032-03449-6_5

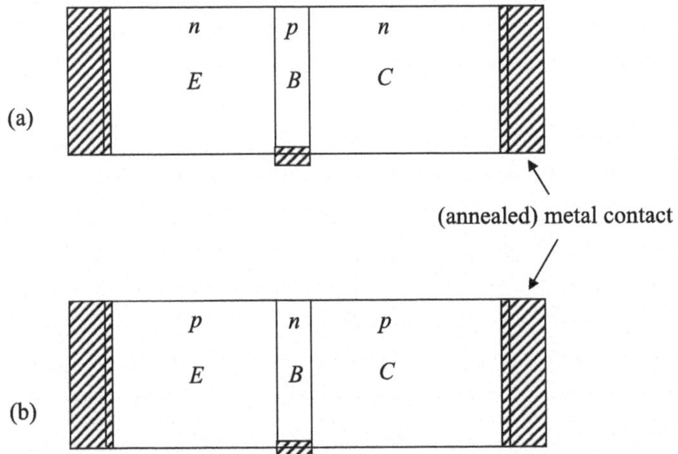

Fig. 5.1 Showing material structure of **a** an npn BJT and of **b** a pnp BJT

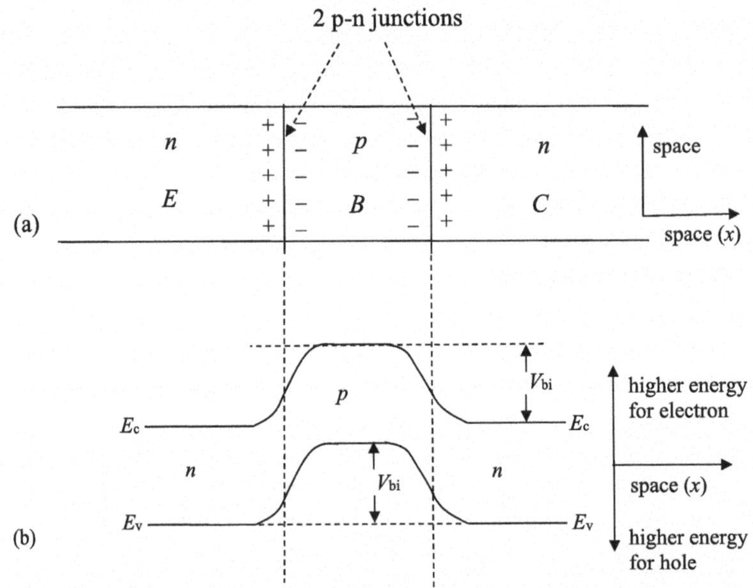

Fig. 5.2 **a** Shows structure and **b** shows band model of BJT for *zero* applied bias voltage. V_{bi} is built-in potential barrier height which is ≈ 0.7 eV for Silicon crystal

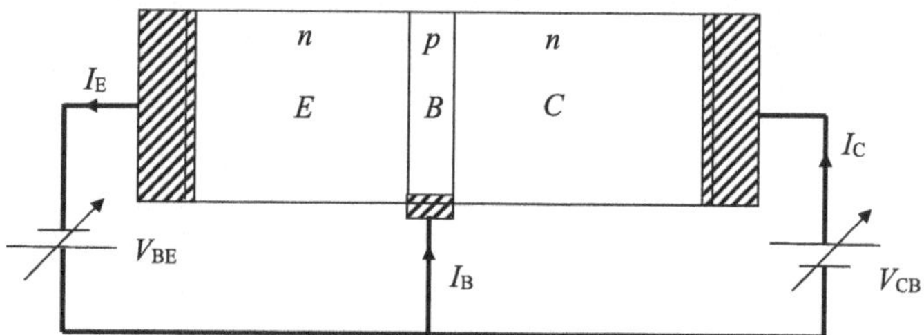

Fig. 5.3 Showing an npn BJT in common base configuration. (Annealed) metal contacts are also shown. Emitter current I_E, collector current I_C and base current I_B are indicated

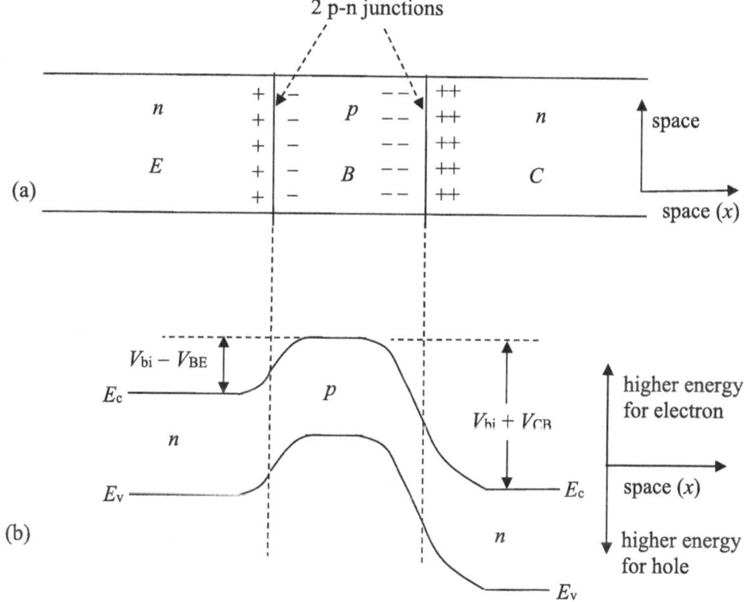

Fig. 5.4 **a** Shows structure and **b** shows band model of BJT for *non-zero* applied bias voltages: forward bias V_{BE} to emitter–base junction and reverse bias V_{CB} to base–collector junction as in Fig. 5.3

5.3 Transistor Currents

See Figs. 5.3 and 5.4; because of forward bias $V_{BE} \geq V_{bi}$, there is a large flux of free electrons from conduction band of emitter to conduction band of base. And there is a small flux of holes from VB of base to VB of emitter. The hole flux is small because base is lightly doped and is thin and hence base contains small number of holes. We have *emitter current* I_E due to the two currents which add to each other.

Because of reverse bias V_{CB}, we expect a small flux of minority carriers of both electrons and holes across the base–collector junction. But in addition to this, the large flux of electrons that has come from emitter to base *slides* through the *downhill* potential profile at base to collector junction and reaches collector constituting what is called collector current I_C. Area of cross section of base–collector junction is large for which resistance of collector terminal is small. Thus in addition to small reverse minority current, we get a large collector current (I_C) which is almost equal to emitter current, $I_C \approx I_E$. This is called *transistor action.*

The collector current I_C is slightly less than the emitter current I_E. Because: (1) some of the electrons that come from emitter to base recombine with holes. These holes are small in number because base is lightly doped and is thin. Due to the recombination, both electrons of conduction band and holes of VB of base are lost from free carriers. (2) Some of the free carriers that come from emitter to base come out through base terminal. This is small because area of cross section of base is small and hence resistance of base terminal is high. We have base current I_B. According to Kirchhoff's current law, $I_E = I_C + I_B$.

5.4 I-V Characteristics of BJT in Common Base (CB) Configuration

As to common base circuit configuration shown in Fig. 5.3, I-V characteristics of BJT obtainable using the circuitry of Fig. 5.5 are as in Figs. 5.6 and 5.7. In Fig. 5.6, we have usual I-V characteristics of a forward biased p–n junction: the emitter to base junction. Figure 5.7 shows the output characteristics: non-zero V_{CB} is not required for non-zero I_C. We have $I_C \approx I_E$ for zero or any non-zero value of V_{CB} if I_E is kept fixed. For negative V_{CB}, both p-n junctions become forward biased. For $V_{CB} > V_{BE}$, collector current I_C

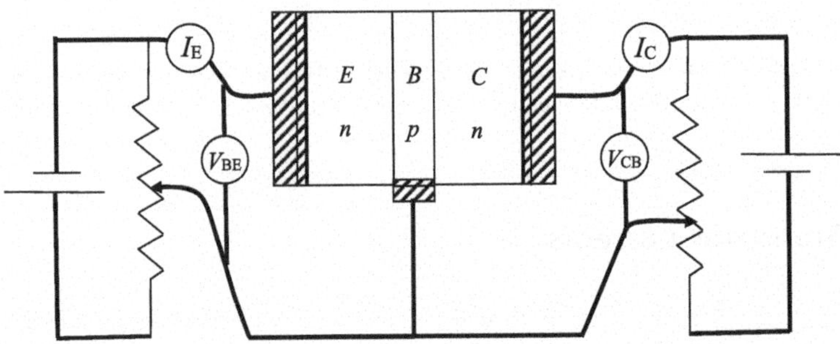

Fig. 5.5 The circuitry that can be used for data collection for I-V characteristics of npn BJT in CB configuration

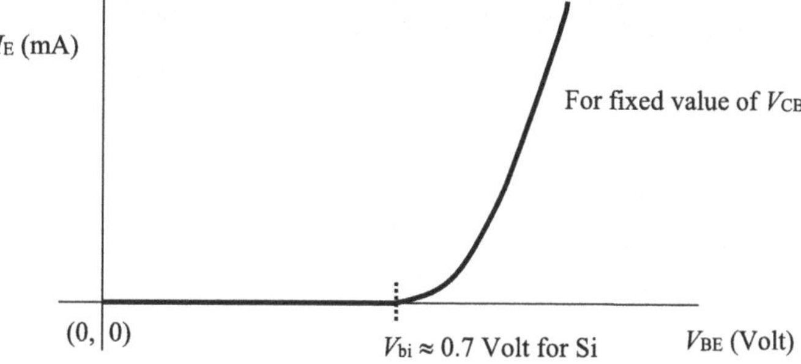

Fig. 5.6 Showing *input* I-V characteristics for the circuit of Fig. 5.3. Emitter current I_E versus emitter–base voltage V_{BE} is shown for fixed value of base–collector voltage V_{CB}

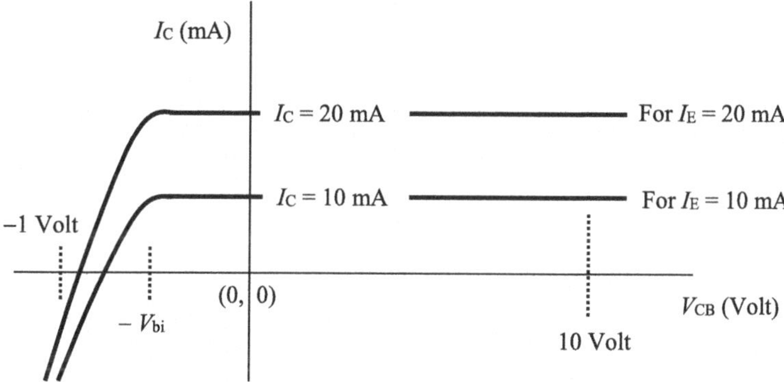

Fig. 5.7 Showing *output* I-V characteristics for the circuit of Fig. 5.3. Collector current I_C versus base–collector voltage V_{CB} are shown for fixed values of emitter current I_E

changes direction. One thing that may be noted is that V_{BE} drops across emitter to base junction *only*, and V_{CB} drops across base to collector junction *only*.

5.5 I-V Characteristics of BJT in Common Emitter (CE) Configuration

As to common emitter circuit configuration shown in Fig. 5.8, I-V characteristics of BJT obtainable using the circuitry of Fig. 5.9 are as in Figs. 5.10 and 5.11. As can be understood from Fig. 5.8, V_{BE} forward biases *both* the p-n junctions simultaneously; but V_{CE} reverse biases base to collector junction *only*; V_{CE} *does not* drop across emitter to base junction *at all*.

In Fig. 5.10, we have input characteristics. We have usual I-V characteristics of a forward biased p–n junction: the emitter to base junction with base current in μA scale. In Fig. 5.11, we have output characteristics. We have saturation region, active region and

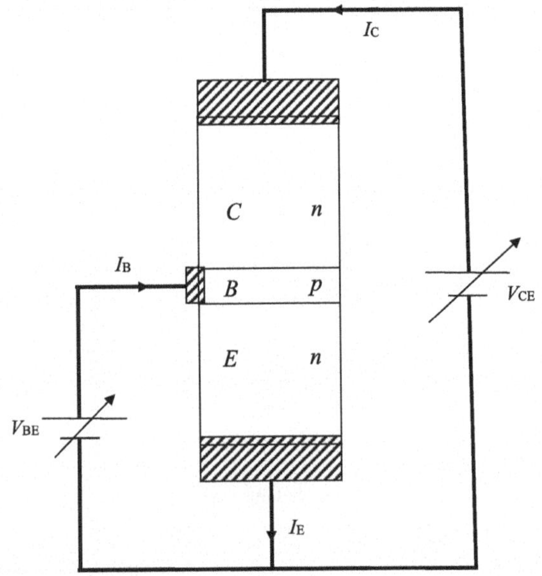

Fig. 5.8 Showing an npn BJT in common emitter (CE) configuration. (Annealed) metal contacts are also shown. Emitter current I_E, collector current I_C and base current I_B are indicated

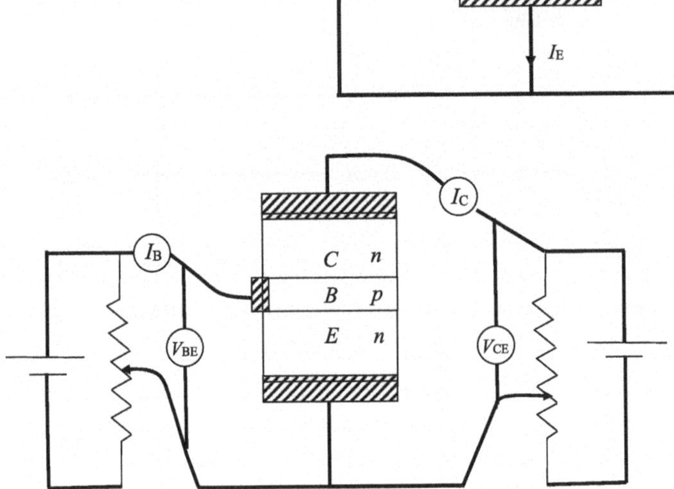

Fig. 5.9 The circuitry that can be used for data collection for I-V characteristics of BJT in CE configuration

Fig. 5.10 Showing input I-V characteristics for the circuit of Fig. 5.8. Base current I_B versus emitter–base voltage V_{BE} is shown for fixed value of emitter–collector voltage V_{CE}

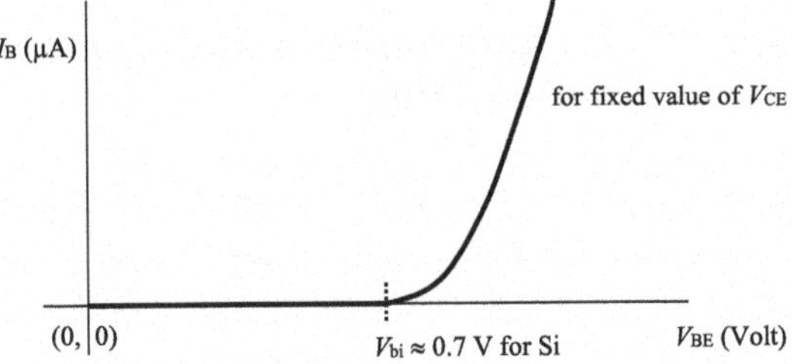

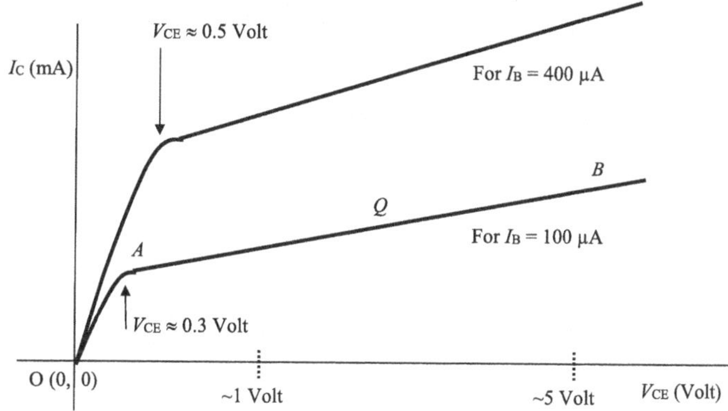

Fig. 5.11 Showing output I-V characteristics for the circuit of Fig. 5.8. Collector current I_C versus emitter–collector voltage V_{CE} curves are shown for fixed values of base current I_B

cut-off region. For small values of V_{CE} in Fig. 5.11, I_C versus V_{CE} curve *OA* is steep; in this interval both p-n junctions get forward biased by V_{BE}; this region is called saturation region. I_C versus V_{CE} curve for $I_B = 0$ is said to be in cut off region. For $V_{CE} >$ around 0.5 V in output I-V curve (*AB* in Fig. 5.11), the BJT is said to be in active region in which reverse biased base–collector junction is highly resistive.

5.6 Biasing npn BJT in CE Configuration

As to BJT in CE configuration, transistor *biasing* is the process of arranging it in a DC circuit so that pair of values $Q(V_{CE}, I_C)$ occurs in active region of its output characteristics. Here Q is operating point; see Fig. 5.11. Voltage divider bias is the best way to bias BJT in CE configuration, see Fig. 5.12.

Looking at Fig. 5.12, since base current is negligible (in mA scale), we have almost same current passing through R_1 and R_2. As such the battery voltage V_{CC} divides into two parts: V_{R1} and V_{R2} across R_1 and R_2 respectively in the ratio R_1: R_2; hence the name voltage divider bias. Here is an approximate analysis of the circuit.

Voltage drop across R_2 is

$$V_{R2} = \frac{R_2}{R_1 + R_2} V_{CC}.$$

Hence voltage drop across R_E is

$$V_{RE} = V_{R2} - V_{BE} = \frac{R_2 V_{CC}}{R_1 + R_2} - V_{bi}$$

where $V_{bi} \approx 0.7$ V for Silicon crystal. As such emitter current

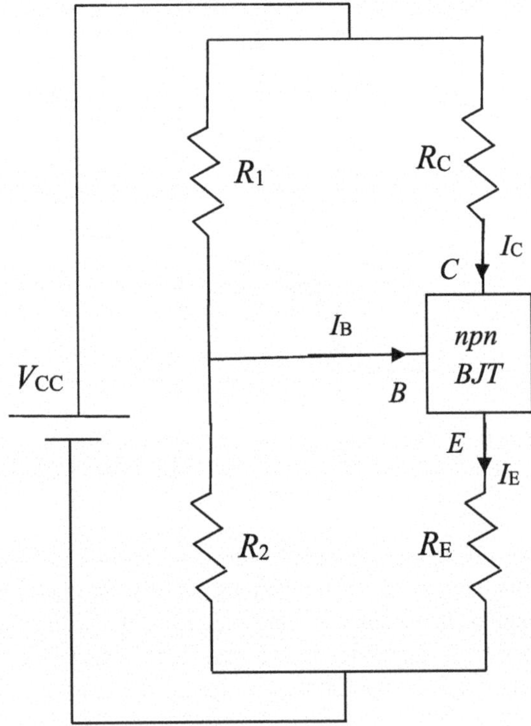

Fig. 5.12 To help describe biasing of a BJT in CE configuration with voltage divider bias

$$I_E = V_{RE}/R_E \approx I_C \tag{5.1}$$

As such

$$V_{CE} = V_{CC} - R_C I_C - R_E I_E = V_{CC} - (R_C + R_E) I_C \tag{5.2}$$

Equations (5.1) and (5.2) determine the pair (V_{CE}, I_C) called operating point Q as indicated in Fig. 5.11 if resistances of the 4 resistors (R_1, R_2, R_C, R_E) and value of V_{CC} are given.

Use of randomly chosen values of R_1, R_2, R_C, R_E and V_{CC} in Eqs. (5.1) and (5.2) may lead to operating point Q away from active region of the output characteristics. Hence in practice, we do the *converse*: we choose a pair of values (V_{CE}, I_C) = (5 V, 10 mA) e.g. in the active region of output characteristics of BJT. With a chosen value of V_{CC}, we then calculate resistances of the 4 resistors and thereby get the desired bias circuit of Fig. 5.12. This is called design of bias circuit. The sequence of steps in this regard is as follows. Quantities of greater interest are in bold face.

(1) With chosen value of $\boldsymbol{V_{CC}}$ **= 20 V** and chosen operating point (V_{CE}, I_C) = (5 V, 10 mA), Eq. (5.2) gives $R_C + R_E$ = 1.5 kΩ. We can take $\boldsymbol{R_C}$ **= 1.2 kΩ** and $\boldsymbol{R_E}$ **= 330 Ω** from available resistors so that the sum is close to 1.5 kΩ.
(2) Now $V_{RE} = R_E\, I_E \approx R_E\, I_C \approx$ (330 Ω)(10 mA) = 3.3 V.

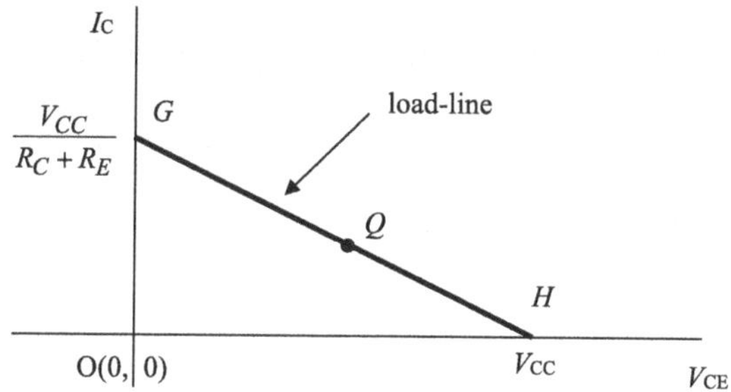

Fig. 5.13 Showing load-line *GH* for voltage divider biased BJT in CE configuration shown in Fig. 5.12. Operating point *Q* lies on this load-line

(3) $V_{R2} = V_{RE} + V_{BE} = (3.3 + 0.7) = 4$ V.

(4) $V_{R1} = V_{CC} - V_{R2} = 20 - 4 = 16$ V.

(5) V_{R1}: V_{R2} = R_1: R_2 = 4: 1. We can choose $\mathbf{R_1 = 4.7\ k\Omega}$ and $\mathbf{R_2 = 1.2\ k\Omega}$ from available resistors so as to maintain the ratio 4: 1 closely.

5.7 Load-Line and Operating Point

Kirchhoff's voltage law applied to output loop of the circuit of Fig. 5.12 for voltage divider biased BJT in CE configuration gives (using Eq. 5.2): $V_{CE} = V_{CC} - R_C I_C - R_E I_E = V_{CC} - (R_C + R_E) I_C$ which can be written as

$$I_C = -\frac{V_{CE}}{R_C + R_E} + \frac{V_{CC}}{R_C + R_E} \tag{5.3}$$

Equation (5.3) reveals that pairs of values $Q(V_{CE}, I_C)$ lie on a straight line *GH* called load line; see Fig. 5.13. The point *Q* is called operating point.

5.8 AC Voltage Amplification by BJT in CE Configuration

Here DC quantities are in upper case, while AC quantities are in lower case. If we somehow apply a small amplitude AC voltage across emitter to base junction of BJT in CE configuration biased in active region of its output I-V characteristics (Fig. 5.12), we get a large amplitude AC voltage as v_{CE} and v_{RC}. This is AC voltage amplification.

See Fig. 5.13; *GH* is load-line on which operating point $Q(V_{CE}, I_C)$ must lie. We use voltage divider bias to get *Q* in active region (of output I-V characteristics, see Fig. 5.11) in which emitter to base junction will be forward biased and base to collector junction will be sufficiently reverse biased. As such the base to collector junction will be highly resistive.

See Fig. 5.14. If we apply a small amplitude input AC voltage v_i as shown, because of AC shorts by capacitors C_B and C_E, v_i drops across R_2 and hence across emitter to base

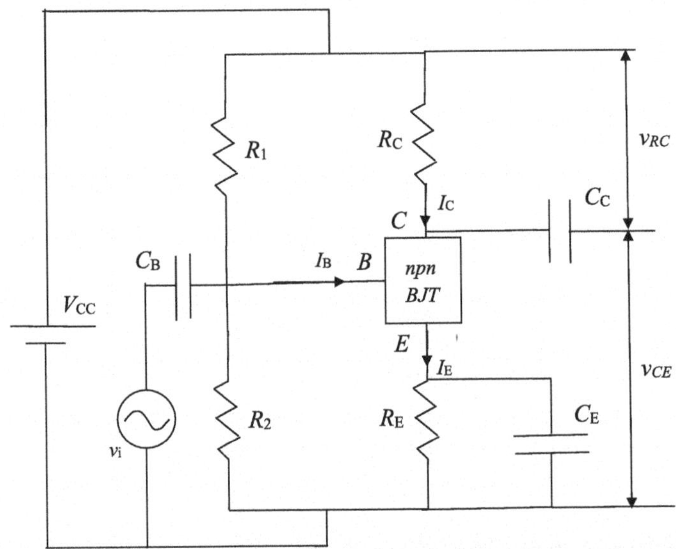

Fig. 5.14 AC voltage amplifier using BJT in common emitter configuration with voltage divider bias. C_B and C_C provide DC isolation of source and load. C_B and C_E provide AC shorts for which v_i drops across R_2 and as v_{BE}

junction. As such there will be a small amplitude variation of V_{BE}. This will result in a large amplitude variation of I_E because small change of V_{BE} causes large change in I_E for adequately forward biased emitter–base junction. See Fig. 5.6. We get large amplitude variation in I_C (because $I_C \approx I_E$) easily passing through downhill potential of highly resistive base to collector junction. Thus we get large amplitude variation of V_{CE}. This is AC voltage amplification.

Operating point Q swings along load-line. See Fig. 5.13. To allow large swing of Q, we use large V_{CC}. To avoid incursion of Q into saturation region, we can use smaller amplitude input AC voltage v_i.

See Fig. 5.13. As Q point moves along the load-line, during positive half cycle of v_i, I_C rises and hence V_{CE} reduces. Hence v_{CE} is 180° out of phase with the input voltage. v_{RC} is in-phase with v_i because I_C and hence V_{RC} both increases in positive half cycle of v_i. Kirchhoff's voltage law applied to output loop gives $v_{CE} + v_{RC} = 0$ or $v_{CE} = - v_{RC}$. As such v_{RC} is also amplified and is in phase with v_i and 180° out of phase with v_{CE}.

Junction Field Effect Transistor (JFET) 6

6.1 Structure of n-Channel Junction Field Effect Transistor (n-JFET)

There is an n-type semiconductor crystal between what are called source S and drain D. There are two p-type regions of the semiconductor called gate G on two sides of the n-type crystal as shown in Fig. 6.1. We have two p-n junctions. We have ohmic metal contacts for source S, drain D and gate G. Depletion regions form even in absence of applied bias voltage. Depletion regions do not contain free charge carriers.

6.2 I-V Characteristics of n-JFET in Common Source Configuration

The circuit configuration of n-JFET shown in Fig. 6.2 is called common source configuration. This is apparent because source S is common to both drain D and gate G. Here V_{DS}, V_{GS} are called drain-source voltage and gate-source voltage respectively. I_D is called drain current. Drain characteristics and so called transfer characteristics are shown in Figs. 6.3 and 6.4 respectively. Gate current is zero in mA scale because of reverse biased p-n junctions. These I-V characteristics can be measured employing the circuitry of Fig. 6.5.

See Fig. 6.3 for drain characteristics. For $0 < V_{DS} < V_P$, we have somewhat ohmic variation of drain current I_D as a function of drain-source voltage V_{DS}. This is because for small values of V_{DS}, volume of depletion regions near the p-n junctions is small and hence source to drain resistance R_{DS} is essentially constant. The region $0 < V_{DS} < V_P$ in Fig. 6.3 is called *ohmic* region.

S. Chowdhury, *Introduction to Electronics*, Synthesis Lectures on Engineering, Science, and Technology, https://doi.org/10.1007/978-3-032-03449-6_6

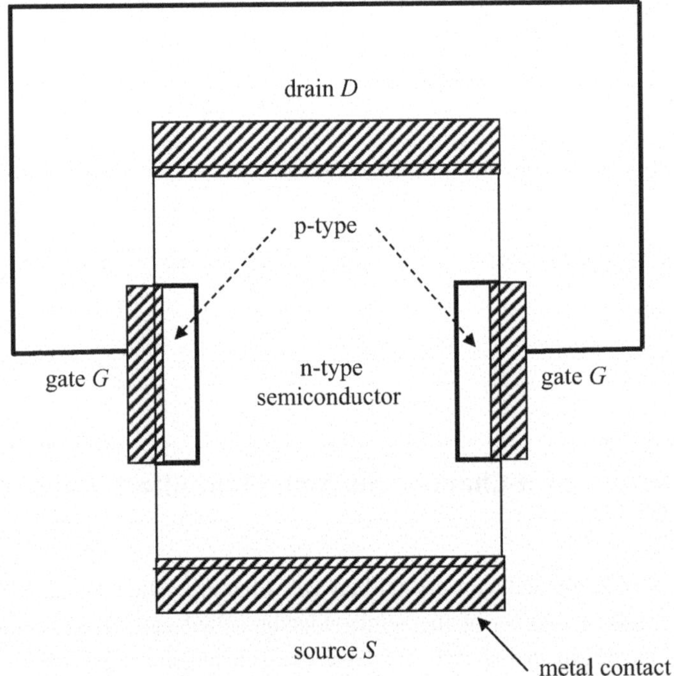

Fig. 6.1 Showing an n-channel junction field effect transistor (n-JFET). (Annealed) metal contacts are also shown

See Figs. 6.2 and 6.3. For values of $V_{DS} > V_P$, volume of depletion region becomes larger and hence R_{DS} increases. At the same time, electrons get more forward thrust to go from source to drain. These two competing effects result in constant drain current I_D though V_{DS} is increased keeping V_{GS} fixed. The region $V_{DS} > V_P$ in Fig. 6.3 is called *active* region.

$V_{DS} = V_P$ is called pinch off voltage indicated in Fig. 6.3. It roughly demarcates ohmic and active regions. $I_D = I_{DSS}$ is the maximum value of drain current for zero gate bias V_{GS}. Notation I_{DSS} comes from drain current with gate shorted to source for which $V_{GS} = 0$.

See Fig. 6.3. For non-zero V_{GS}, pinch-off occurs for smaller V_{DS}. This is because larger values of both V_{GS} and V_{DS} raise volume of depletion region and hence value of R_{DS}. For $V_{GS} = V_P$, we get $I_D = 0$ mA for all values of V_{DS}. Hence V_P is often denoted by $V_{GS\,(OFF)}$.

We now look at transfer characteristics shown in Fig. 6.4. This is I_D versus V_{GS} curve for fixed V_{DS}. It obeys the relation

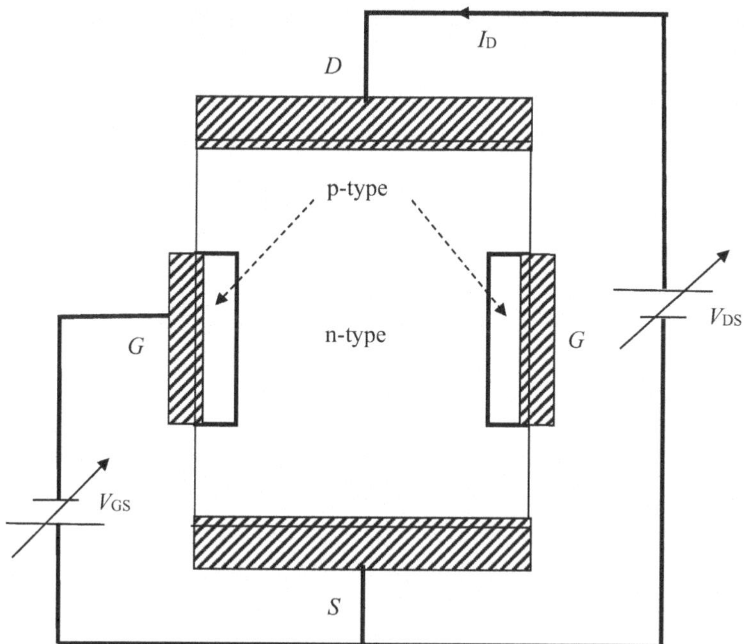

Fig. 6.2 Showing an n-channel junction field effect transistor (n-JFET) in what is called common source configuration. The two gates are connected (which is not shown in the figure for clarity)

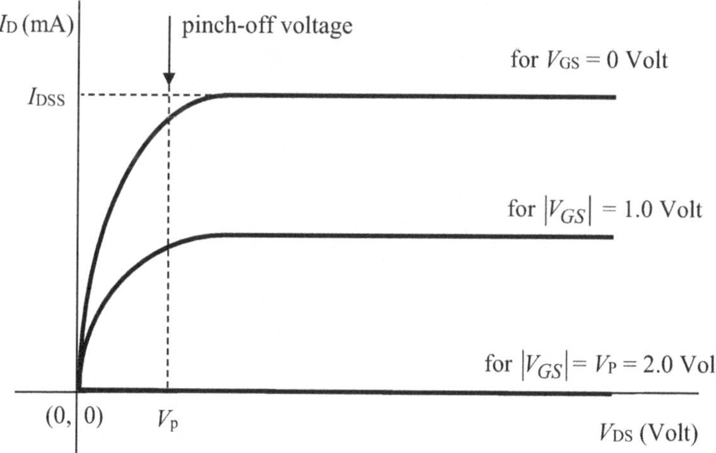

Fig. 6.3 Drain characteristics of n-JFET shown in Fig. 6.2 for common source configuration

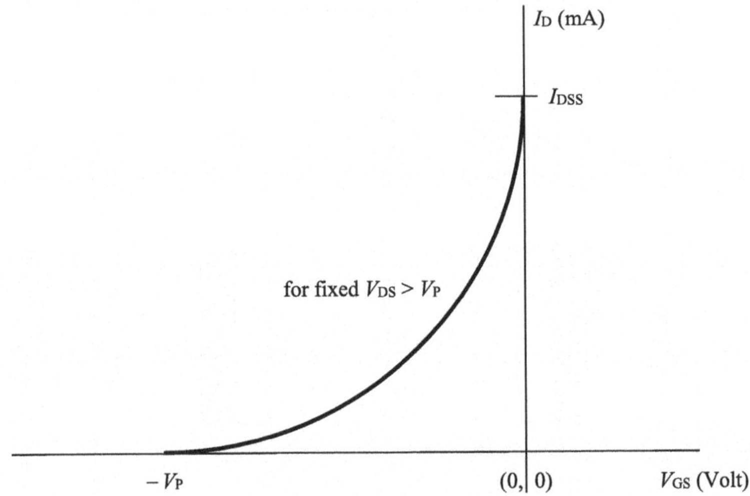

Fig. 6.4 So called transfer characteristics of n-JFET shown in Fig. 6.2 for common source configuration

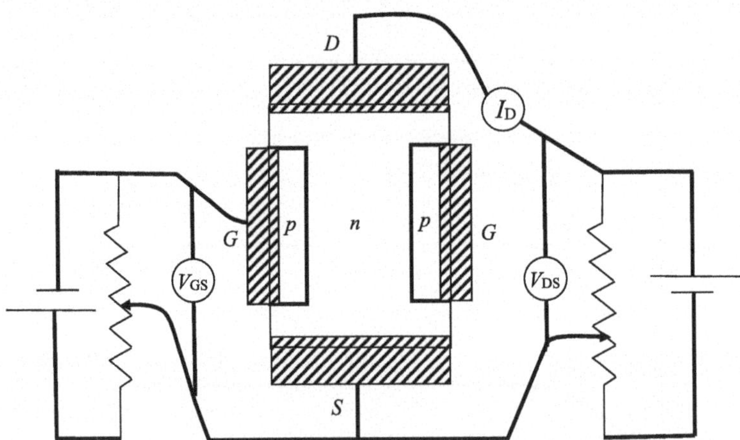

Fig. 6.5 The circuitry that can be used for data collection for I-V characteristics of n-JFET in common source configuration. The two gates are connected (which is not shown in the figure for clarity)

$$I_D = I_{DSS}\left(1 - \frac{|V_{GS}|}{V_P}\right)^2 \tag{6.1}$$

As such for $V_{\text{GS}} = 0$, $0.3V_{\text{P}}$, $0.5V_{\text{P}}$ and V_{P}, we have $I_{\text{D}} = I_{\text{DSS}}$, $0.5I_{\text{DSS}}$, $0.25I_{\text{DSS}}$ and 0 respectively.

6.3 Biasing of and AC Voltage Amplification by n-JFET in Common Source Configuration

Figure 6.6 shows n-JFET in common source configuration with voltage divider bias. Because of reverse bias, gate current $I_G = 0$ mA. As such, for input side, Kirchhoff's voltage law gives $V_{DD} = V_{R1} + V_{R2}$ and we have V_{R1}: $V_{R2} = R_1$: R_2 where V_{R1} and V_{R2} are voltage drops across R_1 and R_2 respectively. As such $V_{R2} = \frac{R_2}{R_1+R_2} V_{DD}$. For input loop, Kirchhoff's voltage law gives $V_{R2} = V_{GS} + V_{RS} = V_{GS} + R_S\, I_D$. Thus we have

$$I_D = -\frac{V_{GS}}{R_S} + \frac{V_{R2}}{R_S} \tag{6.2}$$

Equation (6.2) is called bias line. As Fig. 6.7 shows, intersection of the bias line with transfer characteristics determines allowed value of the pair $Q_1(V_{GS}, I_D)$.

Kirchhoff's voltage law applied to output loop of the circuit of Fig. 6.6 gives

$$V_{DD} = V_{RL} + V_{DS} + V_{RS} = (R_L + R_S)I_D + V_{DS}$$

or,

$$I_D = -\frac{V_{DS}}{R_L + R_S} + \frac{V_{DD}}{R_L + R_S} \tag{6.3}$$

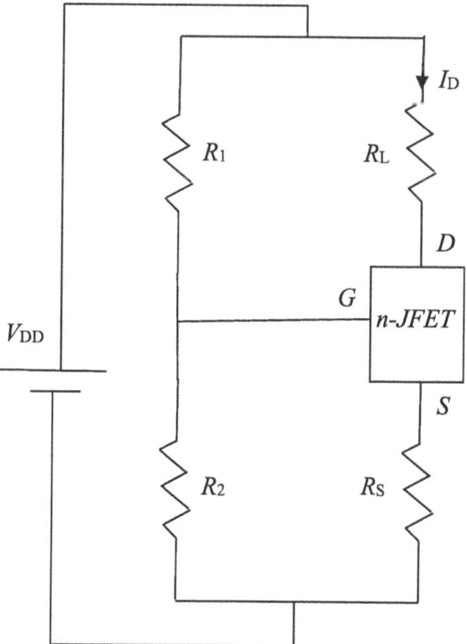

Fig. 6.6 To help describe biasing of n-JFET in common source configuration with voltage divider bias

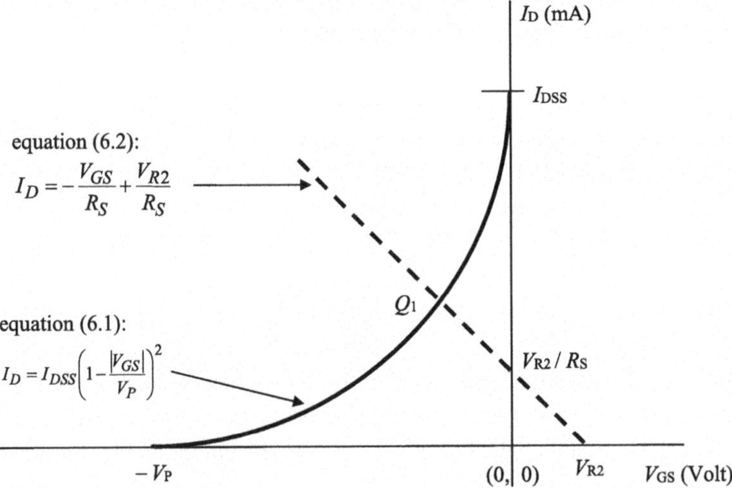

Fig. 6.7 Transfer characteristics of n-JFET and bias line (dashed) plotted together. Q_1 is allowed value of the pair (V_{GS}, I_D)

what is called load line. Allowed values of the pair $Q_2(V_{DS}, I_D)$ can be obtained by plotting the bias line and the load line together as in Fig. 6.8.

We now turn to AC voltage amplification by n-JFET. See Fig. 6.9. Because of small amplitude AC voltage v_{GS}, drain current I_D changes a lot because I_D versus V_{GS} curve is steep; see Fig. 6.8. This means the point Q_1 shifts a lot. Hence Q_2 makes large excursion

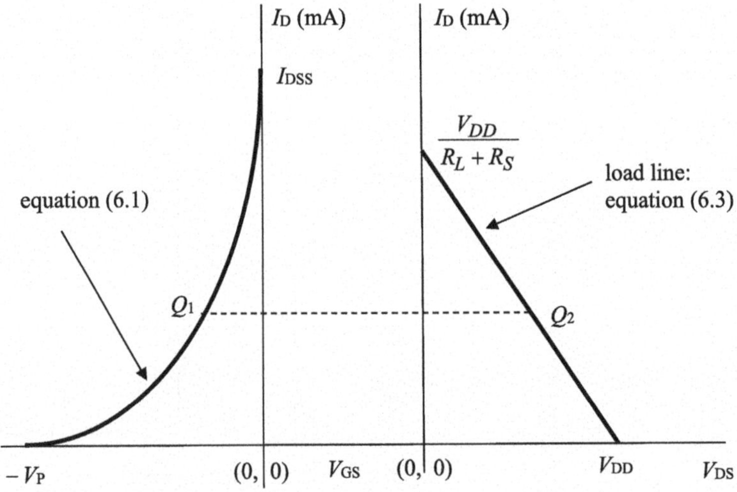

Fig. 6.8 Showing how to locate $Q_2(V_{DS}, I_D)$ using $Q_1(V_{GS}, I_D)$

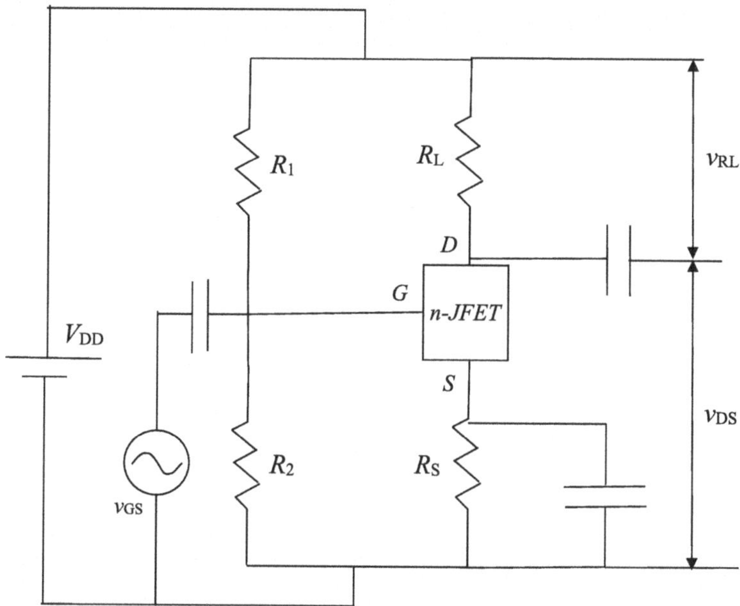

Fig. 6.9 AC voltage amplifier using n-JFET in common source configuration with voltage divider bias

from its zero signal or DC position. Near Q_2, the JFET is in active region having high resistance R_{DS}. Large amplitude oscillatory I_D gets multiplied with the large R_{DS} and hence we get amplified AC output voltage v_{DS} and hence v_{RL} because $v_{DS} + v_{RL} - 0$ as per Kirchhoff's law.

Metal Oxide Semiconductor Field Effect Transistor (MOSFET)

7

7.1 Structure of Depletion Type MOSFET

See Fig. 7.1. The device is called so because of the combination metal–oxide–semiconductor for the gate. The gate is insulated from the semiconductor (Si) by SiO_2 layer which is a very good insulator having large bandgap. Thus no current can flow through the gate terminal. Two n^+-Si regions help metal contacts for source S and drain D become ohmic. An n-Si channel is implanted between source and drain as shown.

7.2 I-V Characteristics of and AC Voltage Amplification by Depletion Type MOSFET in Common Source Configuration

The circuit configuration is shown in Fig. 7.1. Here V_{DS}, V_{GS} are called drain-source voltage and gate-source voltage respectively. I_D is called drain current. Drain characteristics and so called transfer characteristics are shown together in Fig. 7.2. Gate current is zero because of SiO_2 underlying the gate.

Gate bias V_{GS} can be positive or zero or negative. See Fig. 7.2a; for small values of V_{DS}, drain current I_D increases somewhat ohmically before entering active region for larger V_{DS} where I_D is almost constant despite rise of V_{DS}. This is because source to drain resistance R_{DS} is very large. See Fig. 7.2b; I_D versus V_{GS} curve is steep. Hence small change of V_{GS} changes I_D a lot.

This large variation of drain current I_D passing through large R_{DS} causes large variation of V_{DS}. Hence small amplitude AC voltage with V_{GS} causes large amplitude AC voltage with V_{DS}. Thus MOSFET can be used as amplifier.

S. Chowdhury, *Introduction to Electronics*, Synthesis Lectures on Engineering, Science, and Technology, https://doi.org/10.1007/978-3-032-03449-6_7

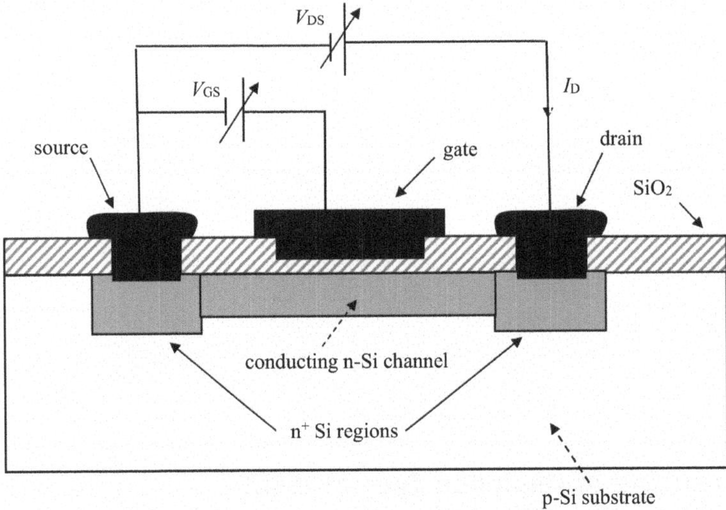

Fig. 7.1 Showing structure and typical circuit connection of a depletion type n-channel metal oxide semiconductor field effect transistor (MOSFET) in common source configuration

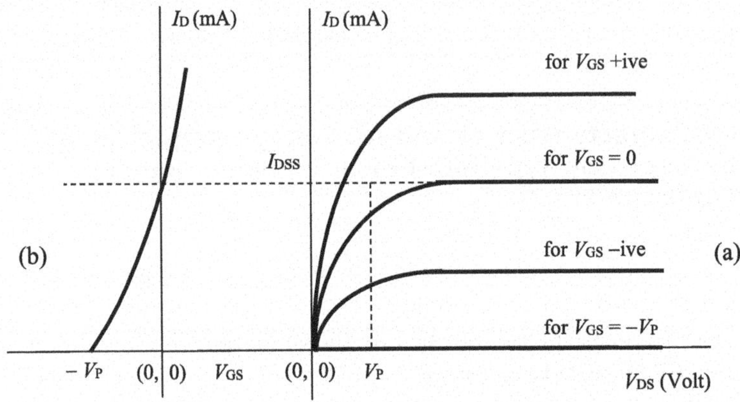

Fig. 7.2 **a** Drain and **b** transfer characteristics of depletion type MOSFET for common source configuration shown in Fig. 7.1. The transfer characteristics obey the equation $I_D = I_{DSS}\left(1 + \frac{V_{GS}}{V_P}\right)^2$

In Fig. 7.2b, if negative value of V_{GS} is used, the MOSFET is in depletion mode and if positive value of V_{GS} is used, the MOSFET is in enhancement mode. Negative gate bias drives electrons away from conducting channel reducing the drain current I_D.

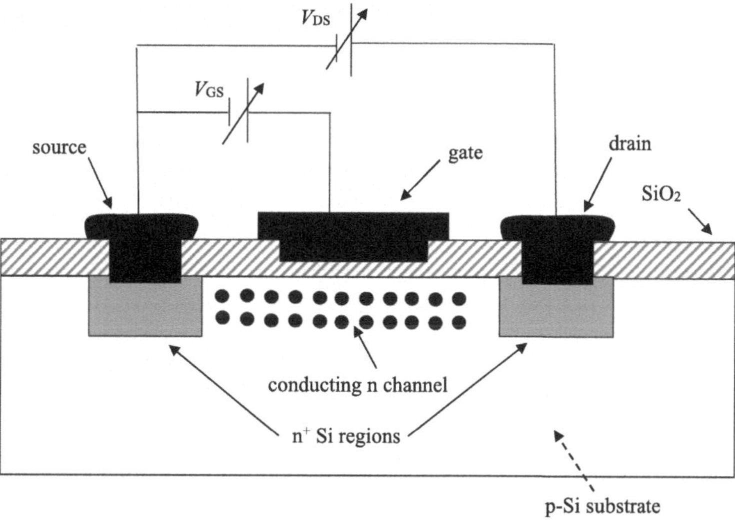

Fig. 7.3 Showing structure and typical circuit connection of an enhancement type n-channel metal oxide semiconductor field effect transistor (MOSFET)

7.3 Structure of Enhancement Type MOSFET

See Fig. 7.3. We have an enhancement type n-channel MOSFET. There is no channel for flow of current between source and drain for zero gate bias. For sufficient positive gate bias, minority electrons of conduction band of p-Si gather between the two n^+-Si regions forming a channel for flow of electrons between source and drain. Minimum positive gate bias needed for formation of the channel is called threshold voltage V_T. For $V_{GS} > V_T$, the channel allows larger drain current I_D. Because of insulating SiO_2 layer, the electrons of the channel do not or cannot go to metal gate.

7.4 I-V Characteristics of and AC Voltage Amplification by Enhancement Type MOSFET in Common Source Configuration

Figure 7.4 shows transfer characteristics of enhancement type n-channel MOSFET. It obeys the equation

$$I_D = k(V_{GS} - V_T)^2 \tag{7.1}$$

where k can be determined using any particular measured pair of values (V_{GSp}, I_{Dp}). Thus

$$k = I_{Dp}/(V_{GSp} - V_T)^2$$

Figure 7.5 shows drain characteristics of enhancement type n-channel MOSFET for common source configuration shown in Fig. 7.3. For small values of V_{DS}, drain current I_D increases somewhat ohmically before entering active region. For larger V_{DS}, drain current I_D is almost independent of V_{DS}. This means source to drain resistance R_{DS} becomes larger.

Because of small amplitude AC voltage added to V_{GS}, drain current I_D changes a lot because I_D versus V_{GS} curve is quite steep, see Fig. 7.4. This large variation of I_D passes through large R_{DS} giving rise to a large amplitude AC voltage v_{DS}. Thus we get AC voltage amplification.

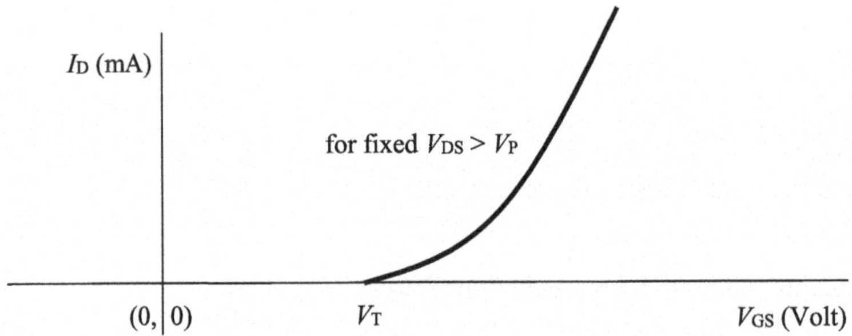

Fig. 7.4 So called transfer characteristics of enhancement type n-channel MOSFET shown in Fig. 7.3 for common source configuration

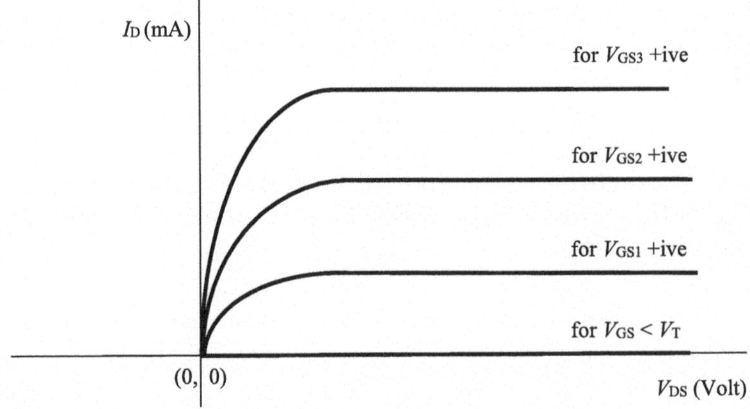

Fig. 7.5 Drain characteristics of enhancement type n-channel MOSFET for common source configuration shown in Fig. 7.3. Here $V_{GS3} > V_{GS2} > V_{GS1}$

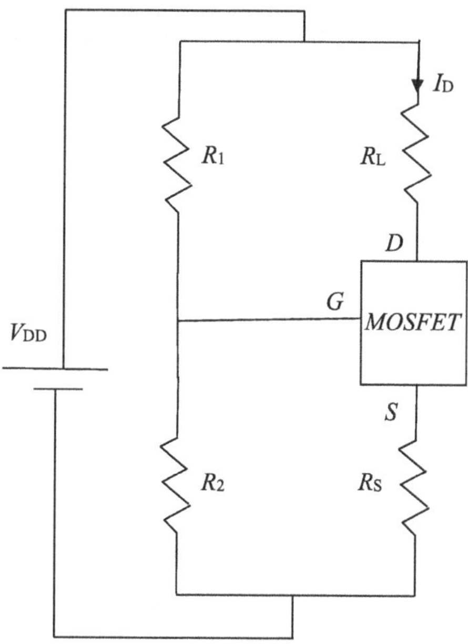

Fig. 7.6 To help describe biasing of MOSFET in common source configuration with voltage divider bias

7.5 Biasing of Depletion Type and Enhancement Type MOSFETs

Figure 7.6 shows MOSFET in common source configuration with voltage divider bias. As in the case with JFET, we have bias line given by

$$I_D = -\frac{V_{GS}}{R_S} + \frac{V_{R2}}{R_S} \tag{7.2}$$

As Fig. 7.7 shows, intersection of the bias line with transfer characteristics determines allowed value of the pair $Q_1(V_{GS}, I_D)$. Depending on value of slope of bias line determined by R_S, V_{GS} at Q_1 can be positive or negative. As such, the MOSFET will operate in enhancement or depletion mode respectively.

Use of $Q_1(V_{GS}, I_D)$ with load line given by

$$I_D = -\frac{V_{DS}}{R_L + R_S} + \frac{V_{DD}}{R_L + R_S} \tag{7.3}$$

determines the pair of values $Q_2(V_{DS}, I_D)$ which we need to get in the active region of output characteristics.

We have bias line given by

$$I_D = -\frac{V_{GS}}{R_S} + \frac{V_{R2}}{R_S} \tag{7.4}$$

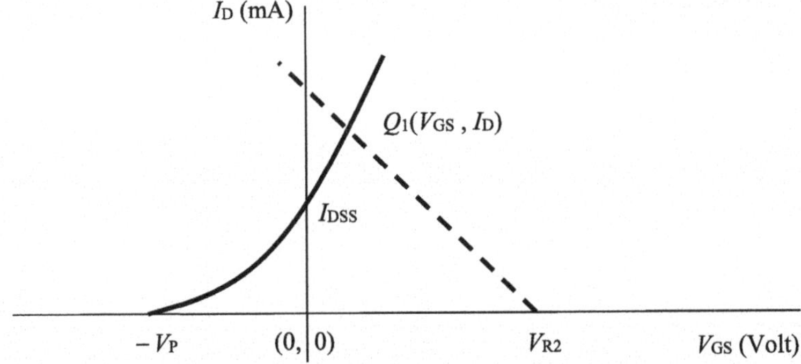

Fig. 7.7 So called transfer characteristics of depletion type n-channel MOSFET shown together with bias line (dashed). For intersection point Q_1, V_{GS} can be positive or negative depending on value of R_S which determines slope of the bias line

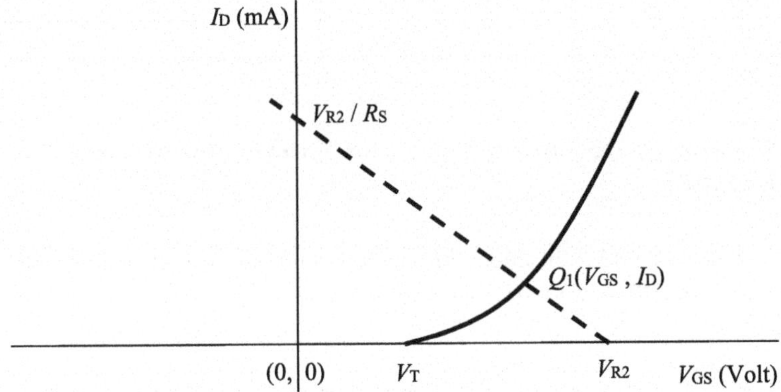

Fig. 7.8 So called transfer characteristics of enhancement type n-channel MOSFET shown together with bias line (dashed)

As to enhancement type MOSFET, as Fig. 7.8 shows, intersection of the bias line with transfer characteristics determines allowed value of the pair $Q_1(V_{GS}, I_D)$. Use of $Q_1(V_{GS}, I_D)$ with load line given by

$$I_D = -\frac{V_{DS}}{R_L + R_S} + \frac{V_{DD}}{R_L + R_S} \tag{7.5}$$

determines the pair of values $Q_2(V_{DS}, I_D)$ which we need to get in active region of output characteristics.

Bibliography[1]

1. D.L. Eggleston, *Basic Electronics for Scientists and Engineers* (Cambridge, 2011)
2. J.J. Brophy, *Basic Electronics for Scientists*, 5th edn. (McGraw Hill, 1990)
3. D.C. Dube, *Electronics: Circuits and Analysis*, 2nd edn. (Narosa, 2013)
4. R.L. Boylestad, L. Nashelsky, *Electronic Devices and Circuit Theory*, 10th edn. (Pearson, 2009)
5. R.L. Boylestad, *Introductory Circuit Analysis*, 5th edn. (Bell and Howell Company, 1994)
6. T.L. Floyd, *Electronics Fundamentals: Circuits, Devices and Applications* (Merrill Publishing Company, 1987)
7. A.P. Malvino, *Electronic Principles*, 3rd and 6th edn. (McGraw Hill, 1984, 1999)
8. K.C.A. Smith, R.E. Alley, *Electrical Circuits: An Introduction* (Cambridge, 1992)

[1] For device Physics on different devices including BJT and JFET, readers are referred e.g. to: Sujaul Chowdhury; Solid State Device Physics; 2nd ed.; American Academic Press (2024).

S. Chowdhury, *Introduction to Electronics*, Synthesis Lectures on Engineering, Science, and Technology, https://doi.org/10.1007/978-3-032-03449-6

The manufacturer's authorised representative in the EU is Springer Nature Customer Service Centre GmbH, Europaplatz 3, 69115 Heidelberg, Germany. If you have any concerns regarding our products, please contact ProductSafety@springernature.com

Printed and bound by CPI Group (UK) Ltd, Croydon, CR0 4YY
10/07/2026
02163613-0001